The Non-Solutions Project
of
Mark Z. Jacobson

How a Stanford professor poisons the well against credible solutions in the climate change debate

**Written
by
Mathijs Beckers**

We either base our 'confidence' on reason (evident probabilities, past experience, competence, etc) or we base our beliefs on faith, which is blind by definition. Faith is the most dishonest position it is possible to have, because it is an assertion of stoic conviction that is assumed without reason and defended against all reason. If you have to believe it on faith, you have no reason to believe it at all. — AronRa

Part one: An introduction to the 100% WWS thesis   1

  The solutions project   1

  Congressional testimony   5

Part two: the shortcomings of the 100% WWS thesis   9

  Ambiguity and inconsistencies   10

  de-acidification & carbon capture   19

  Desalination & Recharging fresh water   30

  Material requirements   36

  Run the numbers   54

Part three: the counterargument   66

  Energy reality   66

  Misconceptions about nuclear energy   73

  Groundbreaking nuclear innovations   89

  Speculations about a future of non-carbon energy   112

Conclusion   125

Appendix   132

  References   132

  About the author   143

Acknowledgements

I want to thank these people for helping me write this book through encouragement, advice, contribution, and review.

Dr. Alexander Cannara

Prof. Ken Caldeira

Prof. Per Peterson

Dr. James Conca

Doctoral Researcher Ben Heard

Colin Boyd (editing)

Robyn Gough (cover design)

Preamble

Mark Z. Jacobson, a Stanford University Professor, has presented a roadmap in which he postulates that it should be possible to decarbonize all energy production and consumption on the world by 2050. This document is called the 100%WWS Roadmap and is available at "the Solutions Project" website. In this book I am going to show you that the feasibility of this roadmap has been weighed incorrectly. Also, Jacobson has shown that he doesn't accept criticism from anyone, scientists or not. He is backed by one of the wealthiest Universities in the world, and he has the outspoken support of celebrities like Leonardo DiCaprio, Bill Nye and Mark Ruffalo. His claims have swayed influential journalists and leading politicians, but his errors make this rebuttal essential. Academic pedigree or expertise becomes irrelevant when the facts don't support the claims. Only evidence—or its absence—matters.

I don't ask, nor expect, you to believe me. I want you to use your mind and determine whether the 100%WWS Roadmap is realistic or not. The 100%WWS Roadmap is based on published and peer-reviewed papers; who am I to question the validity of this roadmap?

The case against the 100%WWS Roadmap will be based on physics and calculations. The counter argument: addressing the need for nuclear energy to be included in a future low-carbon energy mix.

Does the Solutions Project provide a Polaris by which to navigate through our age of perils?

Units

Most units in this book are in Metric, Celsius, and Short-Scale; Kilo = 10^3; Mega = 10^6; Giga = 10^9; Tera = 10^{12}; unless stated otherwise.

What others have to say

In "The Non-Solutions Project", Mathijs Beckers demonstrates that effective strategies to avoid dangerous climate change must depend on well-established facts, not hopeful fantasies. Beckers shares Jacobson's goal of creating an energy system that does not use the sky as a waste dump for carbon dioxide pollution, but documents how Jacobson's fantastical vision, built on a cloud of wishful thinking, evaporates when confronted with the facts. Beckers points to technically-feasible paths forward that would lead us to an energy system that meets human needs while protecting the riches with which nature has endowed us. — **Ken Caldeira, Professor at the Department of Earth System Science at Stanford University**

Mathijs Beckers provides important analysis of 100% WWS scenarios, comparing them with conventional energy systems based upon the materials inputs needed to construct the infrastructure. While it has long been known from life cycle assessment studies that construction of wind and solar generation infrastructure requires much larger quantities of steel, concrete, copper and other commodities than does conventional nuclear and fossil generation infrastructure, Beckers' analysis of the total materials inputs for recent 100% WWS scenarios, particularly copper, is sobering and deserves attention by energy policy experts. — **Per Peterson, Professor at the Nuclear Engineering Department at UC Berkeley**

Mathijs Beckers exposes Jacobson's dream of a 100% renewables mix by mid-century as the fantasy it is, and as the path that will likely prevent us from achieving a real solution to our climate crisis. Ignoring what is actually going in the world is not a solution. — **Jim Conca, Senior Scientist, UFA Ventures, Inc.**

People opposed to nuclear power often make the argument "We don't need nuclear: we can use renewables for a low-carbon future." Chief arguer for "Only renewables needed" is Mark Z. Jacobson, a professor at Stanford. Jacobson heads the "the Solutions Project" which supposedly shows how the world's energy can be supplied with only "WWS" — Wind Water Solar. To combat his numbers, which can most charitably be described as unrealistic, *most of us turn to analysis such as those in the book by Sir David MacKay: Sustainable Energy Without the Hot Air. But then we must do our own calculations, in response to Jacobson's numbers.*

The wonderful thing about science and engineering is that numbers count. You might be a professor at Stanford or you might be the Queen of England. But this buys you nothing: if your numbers are incorrect, they are incorrect. We all owe a debt to Mathijs Beckers for showing us the numbers for WWS, and showing how this non-nuclear, non-fossil plan simply will not work. — **Meredith Joan Angwin, Physical Chemist, former project manager at the Electric Power Research Institute. Author of *Campaigning for Clean Air, Strategies for Pro-Nuclear Advocacy*.**

In "The Non-Solutions Project", Mathijs Beckers provides insights into why the foundation premises of the Wind-Wave-Solar energy proposals are more than just flawed, they are dangerous. Dangerous both for the consequences those pathways deliver to the environment, and for the distraction they create from implementing credible pathways. A strength of the work is that Beckers has written for and about the challenges of the world we will actually live in, rather than backfilling preferred solutions to implausible futures as the WWS project has done. For Beckers there is no shying away from the poverty, water, ocean acidification and geoengineering requirements we will face, all demanding a clean-energy response, that invalidate the foundation of the WWS work every single day. Many of the issues Beckers' identified are consistent with our comprehensive review of 100% renewables studies that includes, but is not limited to, the work of Mark Jacobson. However the wheels of academia move slowly, so I am pleased to see this readable popular response as a contribution to one of the most important discussions in the world today. — **Ben Heard, doctoral researcher, University of Adelaide. Founder and Executive Director, Bright New World.**

Mathijs Beckers' new book is a devastating take-down of the dogmatic insistence that we can and should power the world solely on renewables. Highly recommended. — **Michael Shellenberger, founder and president of Environmental Progress.**

Part one: An introduction to the 100% WWS thesis

The solutions project

You might not be familiar with the *solutions project*, and therefore this primer will be written. We are going to familiarize ourselves with its premise of the *solutions project* and have a look at how Mark Z. Jacobson—a Stanford University professor of Civil and Environmental Engineering—postulates that humanity can implement a future that is exclusively powered by wind, water, and solar power (subsequently called WWS). Through the deployment of selected non-carbon emitting energy sources, we can alleviate stresses on vital natural cycles, mitigate climate and ocean change damage, and reduce emission caused health issues.

The abstract:

"We develop roadmaps for converting the all-purpose energy (electricity, transportation, heating/cooling, industry and agriculture/forestry/fishing) infrastructure of each of 139 countries of the world to ones powered by wind, water, and sunlight (WWS). As of end of 2014, 3.6% of the WWS energy generation capacity needed for a 100% world has already been installed in these countries... The roadmaps envision 80% conversion by 2030 and 100% conversion of all countries by 2050. The transformation reduces 2050 power demand relative to a business-as-usual (BAU) scenario by ~31.4% due to the higher work to energy ratio of WWS electricity over combustion and the elimination of energy for mining, transporting, and processing fuels..."

"The new plus existing nameplate capacity of generators across all 139 countries is ~45.8 TW, which represents only ~0.5% of the technically

possible installed capacity. An additional ~0.94 TW nameplate capacity of CSP, ~5.1 TW of new solar thermal for heat, and 0.007 TW of existing geothermal heat in combination with low-cost storage is needed to balance supply and demand economically. The capital cost of all new generators (50.3 TW nameplate) is ~102 Trillion in 2013 USD..."

Note that the abstract mentions *"nameplate capacity"*, which is somewhat misleading, for capacity alone is insufficient to determine whether you will generate enough to satiate demand. This is a fundamental flaw in practically all energy-communication. What we need to know is how much energy we consume, rated in TWh and contrast it with the amount of energy generated, also rated in TWh. Only then can we create a level playing field, upon which we can judge technologies individually. Three metrics determine the amount of energy a technology can generate in a year's time: 1. Time. The number of hours in a year; 2. The nameplate capacity; 3. The capacity factor. Capacity factor (CF) is the fraction of nameplate (as-built) energy capacity actually delivered over a relevant period (day, year,...) in the real world. If CF is much below 1.0, we rightly question the investment. If effect, 1 – CF times the as-built investment amounts to "stranded assets" or "non-performing investments", in economic terms. For example, only about 10% of a nuclear power plant fails to return on its investment, while more than 60% of the best-performing wind farms fail to return on their investment.

"I will do anything that is basically covered by the law to reduce Berkshire's tax rate." "For example, on wind energy, we get a tax credit if we build a lot of wind farms. That's the only reason to build them. They don't make sense without the tax credit."[1] — Warren Buffet

Any investment involves resource consumption in both manufacturing and deployment of any technology. Low CF energy sources inflate resource consumption and deflate an investment's true value.

It is important to note that the 102 Trillion USD should not be squared off against capacity, but should be squared off against expected generation. So instead of getting the USD/MW metric, we get the USD/MWh metric. To determine the output of the 100%WWS scenario we have to translate his energy mix from a capacity to an expected generation model. This is fairly easy for on page 10 of the document, which describes this path, we can find a complete quantified summary of all devices needed.

Note, the document in question is called: **100% Clean and Renewable Wind, Water, and Sunlight (WWS) all-sector Energy Roadmaps for 139 Countries of the world—April 2016**

The 100%WWS energy mix per 2050 reproduced:

Technology	Name-Plate Capacity in GW	Capacity Factor	Individual Units needed	Total annual yield in TWh
Onshore Wind	6422	32.5	1,284,400	18,296
Offshore Wind	3812	32.5	762,400	10,860
Wave Device	199	11	265,409	192
Geothermal	97	71.7	840	610
Hydropower	1036	35.9	x	3260
Tidal Turbine	31	25	30,093	68
Res. Roof PV	3937	28.6	9.051E+09	9870
Com/Gov Roof PV	4279	28.6	9.837E+09	10,728
Utility Solar PV	24,432	28.6	5.617E+10	61,253
Utility CSP	1574	22.7	15 679	3132
Total	45,819			118,269

Total 5MW wind Turbines	2 Million
Total 435 Watt Panels	75 Billion

Source: **100% Clean and Renewable Wind, Water, and Sunlight (WWS) all-sector Energy Roadmaps for 139 Countries of the world—April 2016**—*Table2, page 4*
Note: I've used the Capacity Factor information provided by the EIA in the "Electric Power Monthly, of May 2016" [2]

As you can see, the brunt of the weight falls on the shoulders of solar and wind with roughly 82,000 and 29,000 TWh, giving us a total of 111,000 TWh, leaving only 7300 TWh for Wave and Tidal power, Geothermal, Hydropower and Concentrated Solar Power. On page 32 of the roadmap, another 10,000 TWh of possible Wave and Tidal power are provisioned.

From a sum-total energy standpoint, it doesn't make sense to differentiate Commercial and Governmental roof PV, Utility PV, and Residential Roof PV in terms of unit size, and that's why I have opted to show you the amount of panels required for each technology, rather than compartmentalizing them in units of units. PV calculations in the roadmap are based on the SunPower E20 435Watt panel.

Additionally, Solar Heat, CSP and Geothermal heat are provisioned for peaking power. Note that he also mentions these as means of storage, but that's basically a misnomer because you don't put any electricity into these devices, they merely offset electricity usage.

Additional CSP	944	22.7	9442	1878
Solar Thermal	4136	28.6	102,068	10,369
Geothermal Heat	70	71.7	x	

With this added capacity the total generation capacity of 100% WWS reaches roughly 130,000 TWh per year by the 2050's. And it would have taken 102 Trillion US Dollars to get this capacity built, plus uncounted acreage and materials consumption taken from the environment we wish to preserve. Note: this is not LCOE (Levelized Costs Of Electricity), which is the metric by which these kinds of economic matters would usually be determined. Capital cost (CAPEX = Capital Expenditure in economic terms) is only a fraction of the total cost of energy production. So expect to pay more than 102 Trillion USD, and expect these costs to rise as cumulative upkeep for all these technologies kick in.

Congressional testimony

The written testimony to the Committee on Energy and Commerce of the United States House of Representatives by Marc Z. Jacobson is available on his personal website. In this testimony he states that *"Researchers at Stanford University and the University of California have developed roadmaps to transition the energy infrastructures of 139 countries, and the 50 United States to 100% clean, renewable infrastructures running on existing technology wind, water and solar (WWS) power for all purposes by 2050, with 80% conversion by 2030."*

Jacobson thereafter, in the 4th point states that *"the main barriers to a conversion are neither technical or economic; rather they are social and political."*

It is my contention, that the main barrier to a conversion to 100% WWS is based on the availability of raw materials, and therefore a technical one. On the other hand, political and social will have been present for decades, and we may note that despite heavy investments in renewable innovation and deployment, renewable energy sources still are unable to replace fossil fuels on a grand scale. In fact, great attempts have been made to make renewable energy sources, particularly wind and solar, reach grid parity in order to make them able to compete economically with coal- and gas-fired power plants. Despite these efforts, we have yet to see real and significant worldwide shifts from fossil fuels to WWS. By Jacobson's own publications we may conclude that this desired shift is far from a reality. In fact, on page 10 of the 100%WWS roadmap, per April 24th, 2016, Jacobson states that only 3.5% of the required capacity has been installed. We may also note that the growth of wind and solar cannot, and will not, remain exponential because cumulative upkeep and subsequent retirements of wind and solar power generators will increase demand for replacements with equal exponentiality, thus driving the demand for [virgin] materials and chemicals upwards and putting tremendous strain on the extraction industry. The growth curve of wind and solar additions will, therefore, most probably curve down again unless we increase mining activities, and with it emissions and unavoidable negative impacts upon the environment. This increased denudation should be allotted to the land-usage of renewables as well.

The 100% WWS roadmap will only work based on the following assumptions: *"The idea is to electrify everything, thereby eliminating combustion (The burning of fuel) as a source of energy, pollution, and inefficiency."* and *"Electrifying everything reduces power demand relative to conventional fuels by ~32% averaged across all energy sectors due to the efficiency of electricity over combustion."*

In terms of transportation he makes the first elemental mistake. While it is true that the battery electric vehicle (BEV) is roughly 60 to 70% more efficient than a vehicle with an internal combustion engine, the fuel cell vehicle (FCV) is not. According to point three of the methodology presented in the testimony: *"For ground transportation, the technologies to be used include battery electric vehicles (BEVs) and hydrogen fuel cell (HFC) vehicles, where the hydrogen is produced from electricity passing through water."*

It is true that using electricity, in general, is more efficient than using combustion, but it does not logically follow that this efficiency is a third. In fact, there is a multitude of scenarios in which electrification alone is not enough and additional conversion steps are required. For instance, Jacobson advocates the use of hydrogen in transportation, this is a process beset with great losses thanks to an unavoidable number of conversion steps that in the end amount to a loss of energy by at least 50%. It even gets worse if you combust the hydrogen as is stated in point 5 *"Energy for high-temperature industrial processes will come from electric arc furnaces, induction furnaces, dielectric heaters, resistance heaters, <u>and some combusted hydrogen</u>."*

One of the most important reasons to stop combusting fuels like oil, coal, gas, and biofuels (such as wood) is the death-print that these fuels carry. And this is one of the areas where I agree with Jacobson. However, he either knowingly or unwittingly presents a roadmap that hinders us from cutting emissions even faster. *"In the United States, we calculate that 100%*

conversion (he wrote conversation...) to WWS will prevent 60,000~65,000 premature mortalities." That is a laudable goal, but that goal could be achieved faster by including rapid nuclear energy deployments to his energy mix. In fact, according to his energy accounting, we would be able to reach his goals much faster. So by discounting nuclear energy, he accepts that we will reach this lower annual death toll later, thus trading lives in order to avoid nuclear power. This is one of the first of Jacobson's inexplicable paradoxes laid bare.

As we continue to read the testimony we read that a 100% WWS conversion creates millions of jobs, but if we would look at this statement critically we should conclude that this shouldn't be the measure of success. In fact, the measure of success should be to do far more, with less, and not the other way around. It is understood that energy prosperity leads to more stable societies in which population growth would be curbed. The 100% WWS roadmap is only concentrated on the question whether we can make do with limited means, whereas we should be looking to improve society and create an environment of innovation and progress, which cannot based on the mass implementation of the least efficient technologies.

The testimony, therefore, may be considered to be misleading, incomplete, and perhaps even immoral. The idea that it is feasible to transition to a 100% WWS future may have given commissioners the wrong impression, and this is dangerous. Luckily we have testimonies by other scientists and engineers who present different hypotheses—because that's what they are, hypotheticals backed up with evidence. The questions are these: "Whose evidence stands?" and "How is the validity of said evidence determined?"

Thanks to the attention it has got from people who write for newspapers and other media, the Solutions Project has gained a lot of popularity. Jacobson doesn't just testify to congressional committees. In fact, most of his work is aimed at influencing the public by being a keynote speaker at events such as TED, or by writing op-eds. Recently, he wrote an op-ed in the Times Union in response to New York State Governor Cuomo's decision to propose the inclusion of Nuclear Energy in the clean energy portfolio. However, these technologies aren't mutually exclusive. They should work together in order to make a quick conversion to a zero-carbon energy future possible. However, we have to be diligent with our resources and thus need to get a concrete picture before we set off implementing all of these technologies in great numbers. Also, by pitting the general public against nuclear power, our ability to combat climate change effectively, by means of eliminating the combustion economy is being damaged. As such, he is a perfect example of the *green paradox*. Consider this: Jacobson feels the need to deploy as much renewable power as possible, but no new nuclear power, even though experience and evidence shows that our fastest and most complete decarbonization has involved the deployment of nuclear technologies.

Part two: the shortcomings of the 100% WWS thesis

Question number 1: What should we be aiming for?

Answer: To create a prosperous civilization by implementing low-carbon energy sources. Note, until we decarbonize mining and transportation, no energy source will be zero-carbon.

The basic premise of the Solutions Project and the underlying 100%WWS thesis is this: mankind should be able to decarbonize all energy production and consumption by implementing a mix of wind, water, and solar power. This energy mix will be supported by various additional schemes such as smart & super grids and storage solutions. It also assumes that we can cut primary energy demand by roughly 2/3rds when contrasted to EIA predictions, and bases this assumption on the fact that most thermal processes that have electrical counterparts are inefficient, and electrifying them will automatically result in a decrease in energy demand thanks to a great increase in efficiency. However, this roadmap is riddled with erroneous assumptions that will be exposed in the coming chapters.

Note: primary energy is the energy (fuel & renewable sources such as wind, water, solar) that is initially inserted into the entire energy system before it is being converted into functional energy through many conversion steps.

We may also note that the roadmap speaks about end-use and differentiates between the way the EIA weighs transportation of fuels, but the focus should be on how much energy we would need in total, the assumption that all resource exploitation requirements would be gone by implementing WWS is overly optimistic, especially when the plan relies on recycling, which in itself is an energy intense process.

Ambiguity and inconsistencies

When we consider the 100%WWS Roadmap critically, we discover several flaws that allow the roadmap to be falsified. A first serious error is the inconsistency between the sum-total energy figures predicted by the EIA, and those used in the 100%WWS Roadmap: *"The transformation reduces 2050 power demand relative to a business-as-usual (BAU) scenario by ~31.4%"*

$$\frac{130,000\ TWh}{(100-31.4)} \times 100 = BAU = 190,000\ TWh$$

How much energy would humanity consume by the 2050's according to the EIA?

*"The International Energy Outlook 2016 (IEO2016) Reference case projects significant growth in worldwide energy demand over the 28-year period from 2012 to 2040. Total world consumption of marketed energy expands from 549 quadrillion British thermal units (Btu) in 2012 to 629 quadrillion Btu in 2020 and to **815 quadrillion Btu in 2040**—a 48% increase from 2012 to 2040. The IEO2016 Reference case assumes known technologies and technological and demographic trends, generally reflects the effects of current policies, and does not anticipate new policies that have not been announced."* [4]

The EIA extrapolates per-capita energy usage and predicts that we will consume roughly 815 Quadrillion (short scale) Btu (British Thermal unit) in 2040, this translates into *roughly* 240,000 TWh. Also note that this is a 2040 scenario, not a 2050 scenario as predicted by the roadmap. How is it possible that the EIA predicts 240,000 TWh, and the roadmap starts with 190,000 TWh (in 2050...) and then cuts another 31.4%, which gives us a total discrepancy of 110,000 TWh. To give you a clearer impression of what that discrepancy looks like, here are the numbers in a graph:

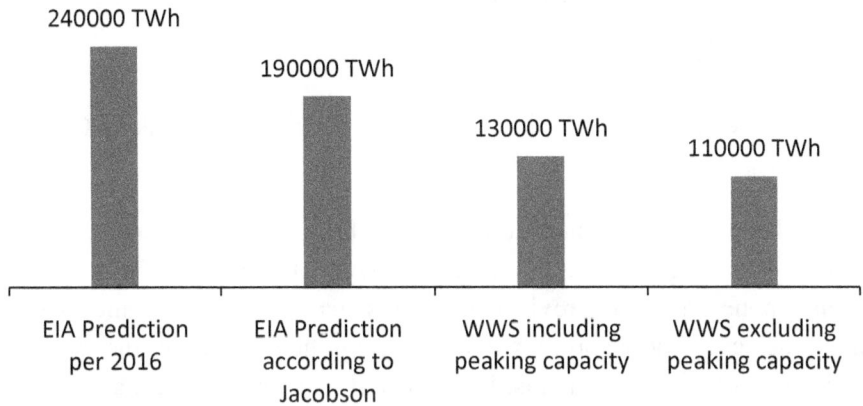

Subsequently, there's a problem with the 102 Trillion USD invested for ~50TW's of capacity. if we want to know how many dollars we'd need to invest to get, for instance, 1MW of WWS capacity, we'd have to do the following.

$$50\ TW = 50{,}000{,}000\ MW$$

$$\frac{102{,}000{,}000{,}000{,}000\ USD}{50{,}000{,}000\ MW} = 2.04\ Million\ USD/MW$$

If we want 50 Terawatts of capacity, we'd have to spend about 2 million USD per Megawatt. But the question is: How much functional energy do you get from this? And how does this translate to actual capacity per dollar?

$$\frac{130{,}000\ TWh}{8766\ Hours\ in\ a\ year} = 14.8\ TW$$

So if we redo prior equation we will find out that the cost per actual capacity, when accounting for capacity factor, is far higher.

$$\frac{102{,}000{,}000{,}000{,}000\ USD}{14{,}800{,}000\ MW} = 6.9\ Million\ USD/MW$$

We can also determine the average capacity factor of the 100%WWS energy mix.

$$\frac{14.8\ TW}{\left(\frac{50\ TW}{100}\right)} = Capacity\ Factor = 29.6\%$$

Almost immediately, we are discovering that the roadmap is likely to be far more expensive than claimed.

The roadmap fails to explain how we will bridge the gap from a carbon-intensive to a carbon-neutral civilization without the use of grid-storage. Because none has been provisioned in this plan, we may assume that we cannot achieve grid-stability without backup generation in the form of *inefficient and leaky* natural gas. Besides, additional power lines alone won't solve the issue, as transmission losses will increase and with it the burden on energy generation. In the roadmap it is framed like this: *"Finally, in the WWS roadmaps, as much additional transmission capacity as possible will be placed along existing pathways but with enhanced lines."*

This is the admission that his plan calls for a significant increase in transmission lines and grid complexity. These transmission lines will not be suspended in thin air; they need to be held up by towers. Even though the footprint of these transmission towers is small, the material footprint is not. This grid enhancement, albeit necessary, will put more stress on material requirements of which we are already pushing the limits.

Additionally, claims are made, for instance on page 3 of the congressional testimony, that conversion to a 100% WWS will provide price stability: *"A 100% conversion will stabilize energy prices because fuel costs of WWS electric power are zero."* This assertion is false if backup power is provided by fueled carbon-burning power plants.

On page 58 of the roadmap document another conflicting statement is made: *"Many uncertainties in the analysis here are captured in broad ranges of energy, health, and climate costs given. However, these ranges miss costs due to limits on supplies caused by wars or political/social opposition to the roadmaps. As such, the estimates should be reviewed periodically."*

This is the paradigm shift that is never mentioned. Prices for WWS technologies won't be determined by fluctuating fuel costs, but by fluctuating costs of raw materials as demand grows, and new [and rich] deposits will probably become harder to find or mine—another fundamental issue omitted. The emphasis will shift from feedstock to material and space availability. Additionally, there are extraction pollution issues in mines and wells, and unintended greenhouse gas leakages, especially where fracking is involved.

To decarbonize civilization, we have to replace carbon burners (primarily combustion engines and generator furnaces/engines) with electrical motors and other electrical conversion systems. The two main proposals in the roadmap are electric engines that run on batteries and/or hydrogen fuel cells. Even though this might not seem apparent, his plan doesn't specify how many vehicles will run on batteries or how many vehicles will run on fuel cells. This is a problem because, if you omit these numbers, a vast discrepancy between demand and generation capacity arises. The idea that we can produce hydrogen from electrolyzing water might seem like a good idea, it is, however, a process beset with energy loss, thanks to the fixed amount of conversion steps. Each conversion step brings about inefficiency, which means that after each step you're left with less energy than you had before.

It is claimed that everything can be done on existing technology, which means that if he mentions BEV's, we may take an example in the Tesla Model S or the Nissan Leaf. If he wants to mitigate the amount of driving, I will take the Nissan Leaf as the standard because it has a battery that is significantly smaller than the Tesla's, and this will give us some leeway as we try to determine whether it is feasible to decarbonize cars by using BEV's.

Consider these simple figures:

> - Current annual worldwide output of batteries: **43 GWh**
> - Projected worldwide output of batteries: **117 GWh / year**
>   *Complete including fully commissioned, partially commissioned, under construction and announced production capacities.* [5]
> - Annual lithium production rate: 32,000 ~ 36,000 tons
>   *Note: there's a deficit, meaning that lithium prices grow, and shortages will increase as demand grows.* [7][8][9][10]
> - Requirement of lithium per KWh: **259 grams**

$$\left(\frac{36,000 \ Metric \ Tons}{100}\right) \times 31 = 11,160 \ Metric \ Tons$$

$$\left(\frac{11{,}160\ Metric\ Tons}{43\ GWh}\right) = 259\ Metric\ Tons\ per\ GWh$$

259 Metric Tons = 259,000,000 Grams

1 GWh = 1,000,000 KWh

259 Grams / KWh

➢ Total known lithium reserves: **13,500,000 tons** [11]

Suppose we can achieve maximum of 117 GWh of annual battery production, and suppose we want to convert half of the world's cars into BEV's i.e. 500 million cars, and we want to achieve this feat by 2050. To do that, we would need to convert roughly 15 million cars per year. Can this be done using 117 GWh worth of batteries? A Nissan Leaf has a battery capacity of 24 KWh. Hypothetically, if we would build 15 million Nissan Leafs per year, we would need 15 million x 24 KWh = 360 million KWh. 360 million KWh equals 360 GWh, which is three times the capacity needed than is currently planned. Then this questions arises: do we produce enough lithium? At 259 grams per KWh, we would need (360 million x 259) / 1 million (conversion from gram to metric ton) = 93 thousand tons of lithium. Which is more than double of what we currently produce. However, one has to consider that the share of lithium consumption allotted to battery production is roughly one third (which is the factor 31 in the calculation on top of this page). This means that lithium production has to be set on a steep incline in order to meet the demand for BEV batteries, while also being able to maintain supply for other processes that depend on lithium.

It's not just lithium that poses a concern, so is the demand for nickel and graphite, which are other essential elements in lithium-ion batteries. Fortunately, there are companies that are trying to revolutionize the battery industry by trying new chemistries and principles. Argonne National Lab has a department which is dedicated to optimizing battery technology. Unfortunately, we're not there yet, and we may conclude that converting regular cars with combustion engines to battery electric vehicles, as of yet, is easier said than done.

FCEV's?

In order to make sense of the idea of creating hydrogen through electrolysis and using it as working energy we have to look at the well-to-wheels chain. We have to specify the steps needed to get a fuel-cell vehicle to work. To make it easy to understand I've made a couple of diagrams. The first shows how much energy is effectively put to use in a fuel-cell vehicle after you've put 100KWh into the chain.

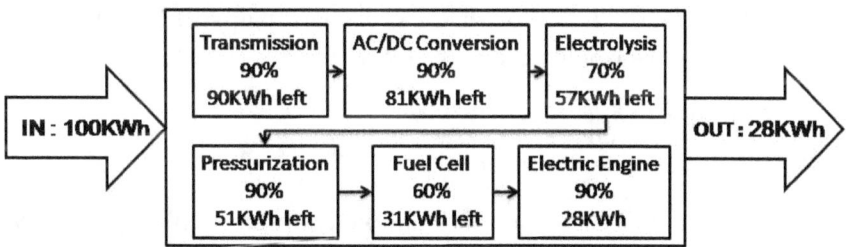

Note: I have given the fuel cell an optimistic 60% efficiency.[12][13]

Look what happens if you choose to burn the hydrogen you produced with electrolysis rather than converting it in a fuel cell.

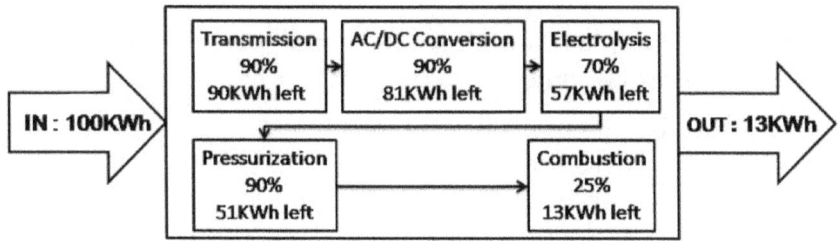

Suppose that, by 2050, half of the cars would be hydrogen fuel-cell cars, the other half is BEVs, and we would have managed to cut driving back so that all of these cars drive an average of 10 thousand kilometers per year, in this case, these cars would drive a combined total of 5 Trillion kilometers. How many kilometers do you get out of your KWh? The Tesla Model S P90D has a range of approximately 500 Kilometers on a full charge. How much energy would be required?

$$\frac{5\ Trillion\ Kilometers}{500\ Kilometers} \times 90\ KWh = 900\ Billion\ KWh$$

Subsequently, normalize to the Terawatt hour metric.

$$900\ Billion\ KWh = 900\ TWh$$

It would take 900 TWh to be able to make 500 million BEVs drive 10 thousand kilometers per year. What about hydrogen fuel cell vehicles? First, they would have an equal end result of 900TWh of working energy since the electric engine and performance of the BEV and the FCEV are the same. Suppose the vehicles' weight, and roll and air resistance are the same. We know that the FCEV wells-to-wheels chain is roughly 28% efficient. This would mean that we would need 3200 TWh:

$$\frac{900\ TWh}{28} \times 100 = 3200\ TWh$$

Combined a billion FCEVs and BEVs would consume 4100 TWh per year, which equals roughly 0.47 TW of 24/7 continuous power. This energy has to come from somewhere. A 5 MW wind turbine produces about 0.013149 TWh in a year:

$$\frac{(((5\ MW\ \times 8766\ hours)/\ 100)\ \times 30CF)}{1\ million} = 0.013149\ TWh$$

This gives us a target, how many wind turbines should we build in order to keep all of these vehicles driving?

$$\frac{4100\ TWh}{0.013149\ TWh} = 311\ thousand\ 5MW\ wind\ turbines$$

Which would mean that 311 thousand of the roughly 2 million wind turbines proposed by the roadmap would need to be built—and periodically rebuilt—for the purpose of (personal) transportation alone. We currently build the equivalent of 10 thousand of these and keep in mind that 2 million are needed, not 311 thousand...

We've seen one possible scenario in which there will as many hydrogen vehicles as BEVs. If we tip the scales in favor of BEVs, we will be straining important commodities required for battery chemistry. On the other hand, if we tip the balance in favor of hydrogen vehicles, we will be increasing energy demand dramatically. Either one will involve an incredibly steep and expensive challenge.

As you can see, a transition from combustion cars to battery electric cars isn't as easy as advertised, and it is doubtful that we can decarbonize all of personal transportation by 2050. We need an additional perspective when reading feasibility roadmaps that are presented to us by academics from highly estimable research institutions.

Even though the number of vehicles hasn't been quantified, it is estimated that ~9% of the total WWS energy will be used for hydrogen production, of which 44.5% will be used for vehicles. This gives the following equation:

$$\frac{130{,}000\ TWh}{100} \times 9 = 11{,}700\ TWh\ for\ hydrogen\ production$$

And

$$\frac{11{,}700\ TWh}{100} \times 44.5 = 5200\ TWh\ for\ vehicles$$

The final inconsistency is the call for electrified flight, for which the developments have hardly surpassed the initial phase—yet feasibility is determined on extant technology. As of yet, we have had a couple of experimental electric prototype aircraft that flew with significant limitations. Consider what Borschberg of the Solar Impulse project said *"We are at the limit of the technology." "You really have to squeeze everything out of what you have."* Solar Impulse2 is the first electric plane to successfully circumnavigate the globe. It is as wide as a Jumbo Jet and seats a grand total of one—the pilot—and has no notable cargo capabilities.

The aeronautic division of NASA is developing several new technologies and innovations to make contemporary airplane combustion engines more efficient, and to make new efficient wings and even complete new hull designs possible. However, the time-to-market of these technologies is relatively long. Also, note that most of these simply improve fuel consumption. The standard for payload efficiency has long been the trans-

Pacific 747 aircraft. The age of electrified flight has yet to begin, will it come in time to facilitate the 100%WWS Roadmap? I doubt it.

de-acidification & carbon capture

I have hoped that we would be able to fix our problems without having to engage in geoengineering. However, the possibility of irreversible positive feedbacks and the possibility of a mass extinction in the oceans have made me reconsider. We desperately need to start working on limiting the acidification of the oceans, for if plankton goes, almost everything else goes (save worms, jellyfish, slugs) and the oceans would end up largely dead, and a dominant source for oxygen and food would dissipate.

I've written a great deal about climate change. If you are a doubter, please go to Bloomberg's website to view an animation which shows you the correlation between greenhouse gases and a warming planet: *"What is warming the Earth?"* It is the simplest representation of the causal factors of anthropogenic climate change.

I will address several issues that are linked to man-made climate change which we could mitigate if we change our ways. These issues have large implications for the 100% Roadmap because there is no accounting for these issues present in his plans.

So what is going on?

For millions of years, the Earth's atmosphere has varied in composition. The earliest probable change that made life as we know it possible was the emergence of microscopic photosynthetic life forms that began to emit oxygen into the oceans as a wasteproduct of their metabolism. Eons passed, and the earliest life forms evolved into the beautiful variety of life we know today. All of this possible thanks to these elements: carbon, hydrogen, oxygen, phosphorous, and sulfur. Carbon atoms are the basis for this process as they are tetravalent, which means that they have four electrons to share. this allows them to form four bonds with other atoms, which is a prerequisite for the formation of long molecules like RNA and DNA, which are the basis for life.

Metabolism is an essential factor in the carbon cycle. As life forms ingest fuel, chemical processes in their bodies transform these fuels into the energy and building blocks that comprise us and all the other life forms on Earth. As a consequence, we're left with some waste products. Two of these waste products are carbon dioxide (CO_2) and oxygen (O_2), enter the carbon cycle. A simplified version would look like this:

A photosynthetic life form uses CO_2, H_2O, and sunlight to create O_2 (as a waste product) and biomass (such as calcified shells); The photosynthetic life form dies and sinks to the bottom of the ocean where it is sequestered underneath layers of sediment; pressure and/or heat transform its carbon content into long chains of hydrocarbons like oil; Normally these hydrocarbons remain in the soil for millions of years while slowly moving into the earth's core thanks to plate tectonics; When back in the mantle the hydrocarbons get "*burned/oxidized*" leaving mostly CO_2; this CO_2 eventually ends up in the atmosphere through volcanic activity, after which the Carbon cycle begins again. However, that cycle has become unbalanced due to the premature extraction of oil and other fossil fuels by humans.

The following describes the dominant CO_2-sequestration system on our planet. Calcifying sea life takes carbonate ions from seawater to build shells and skeletons, and when dying, takes that carbonate to seafloor sediments and eventually turns into to limestone. These animals permanently sequester about 1 billion tons of CO_2 per year—nothing matches their work. See *AAAS Science, Canfield & Kump, vol 339, p533, 2/1/2013*

We're all part of the carbon cycle because we consume hydrocarbons and emit CO_2. Trees and plants are also part of the carbon cycle because they can trap and sequester CO_2 as well. The carbon content of trees and plants usually get released back into the atmosphere when they die. However, when trees and plants die and get submerged (predominantly in wetlands, swamps, and the suchlike) and get covered by sediments, they can be turned into coal over millions of years or they can be turned into methane by microscopic organisms, called Methanogens, that consume biomass.

Now, the carbon cycle has been upset because we are extracting these hydrocarbons and turning them into energy and carbon dioxide millions of years ahead of schedule. Somewhere between a third and half of the carbon dioxide we created has been absorbed by the oceans, where it reacts with water, creating carbonic acid—H_2CO_3—which makes seawater more acidic (less alkaline or basic).

Two reactions that determine how many protons (H^+) are released per molecule of carbon dioxide as it reacts with water to create carbonic acid.

$$H_2CO_3 \rightleftharpoons HCO_3^- + H^+ \text{ and } HCO_3^- \rightleftharpoons CO_3^{2-} + H^+$$

We have evidence to suggest that this vast amount of excess carbon dioxide has lowered the pH by one tenth (note that this scale is logarithmic). Another concern is the increased availability of protons to form bonds with the CO_3^{2-} (carbonate) ions, and this is a problem for shell-forming organisms as it means that there is an unbalance between carbonate ions and Ca^{2+} and Mg^{2+} ions which the shell forming organisms need to form calcium/magnesium carbonate, because they are the main building blocks for shells and bones, and therefore essential for life forms such plankton, oysters, clams, krill, whales etc..[14][15][16][17]

Would it be enough to remove carbon dioxide? According to a paper called *"Long-term response of oceans to CO_2 removal from the atmosphere"*[19], it would not be enough, especially if we keep going on with business as usual. In fact, if we keep introducing excess/man-made carbon dioxide to the atmosphere it will damage marine life, regardless of our efforts to capture and sequester carbon dioxide.

Marine life, including corals have already been damaged. However, bleaching of corals also occurs when waters become too warm, or when they are being exposed to more sunlight[18]. Ocean acidification keeps coral reefs from growing, as less (or no more) calcium/magnesium carbonate is available for reef creation—which is a byproduct of a biochemical process of tiny single-celled symbiotic life forms called Zooxanthellae.[20]

This means that we have to focus on the well-being of marine life, in particular on coral reefs and plankton, to keep the marine life pyramid from collapsing. To accomplish this, we must greatly curtail carbon emissions and start capturing and permanently sequestering carbon dioxide.

If we fail to address these issues, the efforts of those who would implement the 100%WWS roadmap will be in vain, especially when the coral reefs and various calcifying species die because we failed to remediate acidification and warming in time.

Please see marine chemist Andrew Dickson's informative, YouTube presentation called: *"Acidic Oceans: Why Should We Care?—Perspectives on Ocean Science"*

According to the NOAA the atmospheric concentration of CO_2, per august 5th, 2016, is 404.39 ppm. Last year, on the same day, the concentration was 401.31 ppm. Here's a small part of the world-famous Keeling Curve.[21]

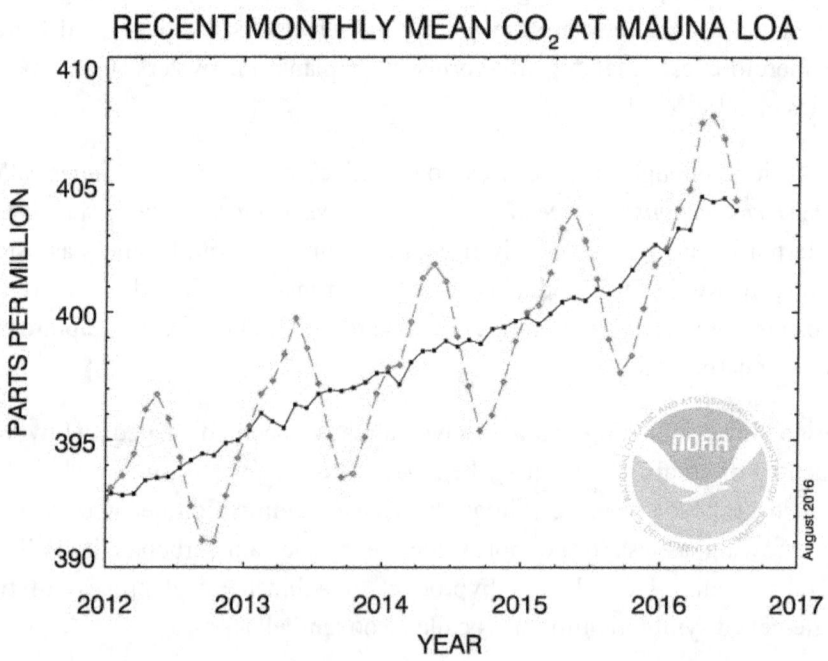

The periodic up and downswings in the concentration levels are representations of the natural carbon cycle, in which the Northern Hemisphere has a large influence due to the carbon dioxide released from forests in autumn and winter, and the uptake during spring and summer. The upward trend, however, is caused by emissions from human sources which do not follow the same seasonal trend. This is tell-tale characteristic is almost linear, and it coincides with our increasing consumption of fossil fuels. Small discrepancies in the linear fashion of the mean line can probably be attributed to economic circumstances that influence the consumption of fossil fuels on a large scale.

Consider these facts and figures from the Fifth Assessment Report of the IPCC[22].

"Annual CO2 emissions from fossil fuel combustion and cement production were 8.3 [7.6 to 9.0] GtC yr–1 averaged over 2002–2011 (high confidence) and were 9.5 [8.7 to 10.3] GtC yr–1 in 2011, 54% above the 1990 level. Annual net CO2 emissions from anthropogenic land use change were 0.9 [0.1 to 1.7] GtC yr–1 on average during 2002 to 2011 (medium confidence)."

"From 1750 to 2011, CO2 emissions from fossil fuel combustion and cement production have released 375 [345 to 405] GtC to the atmosphere, while deforestation and other land use change are estimated to have released 180 [100 to 260] GtC. This results in cumulative anthropogenic emissions of 555 [470 to 640] GtC."

"Of these cumulative anthropogenic CO2 emissions, 240 [230 to 250] GtC have accumulated in the atmosphere, 155 [125 to 185] GtC have been taken up by the ocean and 160 [70 to 250] GtC have accumulated in natural terrestrial ecosystems (i.e., the cumulative residual land sink)."

"Ocean acidification is quantified by decreases in pH. The pH of ocean surface water has decreased by 0.1 since the beginning of the industrial era (high confidence), corresponding to a 26% increase in hydrogen ion concentration".

Note: the term GtC means metric Gigaton Carbon, which is not equal to $GtCO_2$. One GtC equals 3.667 $GtCO_2$.

Roughly 1/3rd of our unnatural CO_2 emissions has dissolved in seas, but even that amount has been sufficient to lower their pH more rapidly than at any time since the great Permian Extinction. Even if we stop all CO_2 emissions today, it would not restore ocean chemistry to pre-industrial levels.

Scenario	Cumulative CO_2 Emissions 2012 to 2100[a]			
	GtC		$GtCO_2$	
	Mean	Range	Mean	Range
RCP2.6	270	140 to 410	990	510 to 1505
RCP4.5	780	595 to 1005	2860	2180 to 3690
RCP6.0	1060	840 to 1250	3885	3080 to 4585
RCP8.5	1685	1415 to 1910	6180	5185 to 7005

As you can see, the IPCC estimates that, given their best climate change mitigation model, cumulative CO_2 emissions from 2012 to 2100 will be at least 510 Gigatons of CO_2, and this does not include the ~2035 Gigatons of CO_2 we have emitted since 1750.

According to the IPCC ocean acidification will look as follows according to these RCP Models:

"Earth System Models project a global increase in ocean acidification for all RCP scenarios. The corresponding decrease in surface ocean pH by the end of 21st century is in the range of 0.06 to 0.07 for RCP2.6, 0.14 to 0.15 for RCP4.5, 0.20 to 0.21 for RCP6.0, and 0.30 to 0.32 for RCP8.5 (see Figures SPM.7 and SPM.8)."

"pH is a measure of acidity using a logarithmic scale: a pH decrease of 1 unit corresponds to a 10-fold increase in hydrogen ion concentration, or acidity."

The corresponding figures:

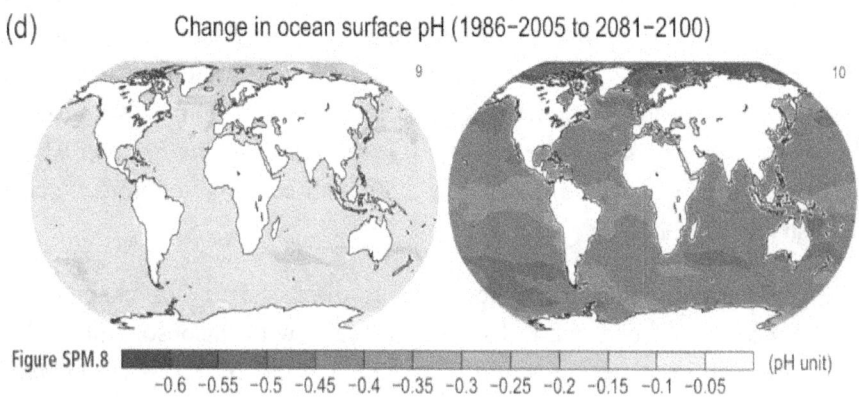

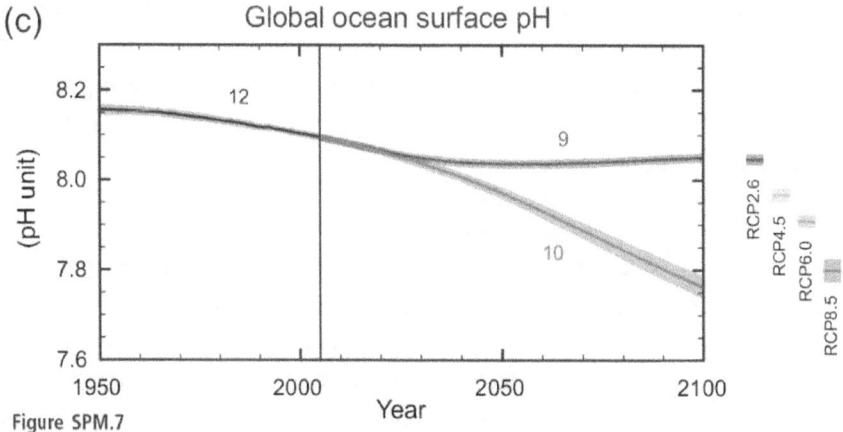

The blue line, which corresponds to the RCP2.6 carbon concentration model, would appear to allow acidity levels to balance out. However, if we consider the business as usual trend, it would dip below a pH of 8.0 by around 2050, which could spell disaster for marine life.

According to the Smithsonian Institute: *"If the amount of carbon dioxide in the atmosphere stabilizes, eventually buffering (or neutralizing) will occur and pH will return to normal."* However, they also state: *"But this time, pH is dropping too quickly. Buffering will take thousands of years, which is way too long a period of time for the ocean organisms affected now and in the near future."* [23]

It should be clear by now that we must ring the alarm bells on ocean acidification and start thinking about reducing the amount of CO_2 in our atmosphere and oceans to a sustainable level, as well as for the rest of the biosphere. It is said that a CO_2 concentration of 350 ppm is sustainable. Please see *"Target atmospheric CO_2: Where should humanity aim?"*[24] to learn why we must try to get below 350 ppm and try to protect ocean chemistry.

One way to return to sustainable levels is discussed in this chapter: *the decarbonization of energy*. I disagree strongly with Jacobson and other academics who promote the *solutions project* and the *100%WWS Roadmap*. The methods proposed in their roadmap are insufficient. We should do far more, as the decarbonization curve that the roadmap provides will not save essential maritime species from extinction. In fact, we need to address the existing carbon-debt first with geoengineering. Several ideas could be successful, But I am going to pick one: the capture of atmospheric CO_2 and sequestering it permanently.

Dr. Alexander Cannara, an engineer who has researched this issue, proposes that we capture carbon dioxide, dissolve it in water, and pump it into porous basalt in order to transform the calcium & magnesium oxides (CaO & MgO), lime, which is a mineral present in basalt[25], and the carbon dioxide (CO_2) to calcium & magnesium carbonate ($CaCO_3$ & $MgCO_3$)[26][27] which is a solid we know as limestone. There are also other minerals present in basalt which can form bonds with dissolved CO_2.

Researchers at BSCP and Carbfix have reported great success in pumping CO_2 into basalt formations for permanent, mineralized sequestration (AAAS Science, 10 June 2016, p1312).

So how do we get from ~405 ppm to 350 ppm of CO_2? Around 1750, at the start of the industrial age, the atmospheric concentration of CO_2 was roughly 280 ppm, and we've added roughly 124 parts per million. We've quantified the amount of man-made CO_2 to be roughly 2035 Gigaton of CO_2. So I think that we need to remove between 850 and 900 $GtCO_2$ to get back to the 350 ppm level. We also we need to account for future emissions, as our combustion economy won't stop without other sources of power being made available.

$$\left(\frac{2035\ Gt}{124}\right) \times (404 - 350) = 886\ Gt$$

How much energy would we need to get all of this CO_2 captured and sequestered?

The molar mass of CO_2 is 44.01 gram/mol. According to an article in the MIT Technology review, we can capture CO_2 with as little as 45 Kilojoules per Mol[28][29].

$45\ Kilojoules\ per\ mol = 12.5\ Watt\ hours\ per\ 44.01\ gram$

So, how much energy do we need if we wanted to capture 900 Gigatons of CO_2?

$900\ Gigatons = 900,000,000,000,000,000\ gram$

$$\frac{900,000,000,000,000,000\ gram}{44.01\ gram} = 204,500,000,000,000\ mol$$

$204,500,000,000,000 \times 12.5 = 2,600,000,000,000,000\ Wh$

$2,600,000,000,000,000\ Wh = 2600\ TWh$

These figures, however, are based on experimental technologies, and they range from 45 kJ/mol to roughly 100 kJ/mol. Also, note that this calculation doesn't include energy needed for mixing and pumping water and CO_2 into the basalt formations. a Harvard study estimated that a 760 meter by 760 meters (0.6 km²) sodium hydroxide spray facility could capture as much as 1 million tons of CO_2 per year, but how many of these can we build? can we stack them? 1 million tons is 0.001 Gigaton. if we would build just one of these, we'd be capturing CO_2 for 900 thousand years, and it would consume 0.003TWh per year.

Suppose we want to remove 900 Gigaton of CO_2 before the year 2050.

$$\frac{900}{(2050-2016)} = 26 \ Gigaton \ per \ year$$

$$\frac{26 \ Gigaton}{0.001 \ Gigaton} = 26{,}000 \ facilies$$

$$26{,}000 \times 0.003 \ TWh = 78 \ TWh \ per \ year$$

If we build 26,000 of these facilities, we could remove roughly 900 gigatons by 2100. The additional 78 TWh per year doesn't seem that incredibly steep. However, full-scale carbon capture facilities might be less efficient.

Another consideration is the area footprint. 26 thousand of these carbon capture facilities would require at least 15,600 km^2 which is about 10% of the surface area of the state of Illinois. There are other technologies that might also work, perhaps even with lower energy consumption and a smaller footprint in materials and area, but we better start capturing CO_2 as soon as possible. Research and innovation should provide means to capture and sequester CO_2 more cheaply and efficiently in the future.

The Global CCS Institute (CCS = carbon capture and storage) estimates that current carbon capture capacity existing and under construction has a total capacity of 68.5 million ton per year (0.0685 Gigaton per year).[30][31] Most of these, however, are tied to carbon emitting industries and are only meant to mitigate their carbon footprint.

The following has been written by Dr. Alexander Cannara and it provides an additional and possibly far more efficient means of mitigating ocean acidification and the harrowing changes that come with it.

"The goals of removing CO_2 emissions, past & present, and of protecting ocean chemistry, can be coupled quite simply—limestone sourcing of lime to distribute into oceans to maintain pH at safe levels, and sequestration of the limestone's CO_2 emitted in making lime. This yields an efficient use of energy and materials, since limestone is about equally, ton-for-ton, composed of CO_2 and lime. This is what cement plants do—kiln limestone to yield lime for cement, but allow the CO_2 to be released, not sequestered. We must stop that, whether the lime goes to protect seas or to make concrete.

*In this way, we sequester ancient CO_2 and protect the oceans' life forms that sequester CO_2 at a far higher rate than any other planetary system can, including our own designs. The cost? We emit >30 billion tons of new CO_2 per year. The oceans dissolve some. If we were to simply try to keep up with that, it would mean making & distributing ~10 billion tons of lime per year, at an energy cost of about 400 kWh/ton. This would require the equivalent of about **900 new, 1 Gwe nuclear reactors**. And, it would require ocean shipping capable of handling about 10,000 tons of lime for distribution over each of the ~1 million oceanic transits/year. We would gradually make inroads on CO_2 in air and acidification in seas—gradually.*

It's just doable today if we get cracking. It would have been far easier if we'd indeed followed John F. Kennedy's lead and eliminated combustion power by about 2000."

$(900 \times ((8766/100) \times 90))/1000 = 7100 \, TWh$

Which, as you can see, would add yet another 7100 TWh to our already increasing energy deficit—if we are serious about mitigating ocean acidification

Desalination & Recharging fresh water

Although the 100%WWS Roadmap speaks about preventing air pollution mortalities and mitigating costs, it doesn't consider increasing water scarcity in much of the world. First, we will look at the costs mentioned, which will be mitigated by implementing his roadmap:

"Costs of climate change include coastal flood and real estate damage costs, agricultural loss costs, energy-sector costs, water costs, health costs due to heat stress and heat stroke, influenza and malaria costs, famine costs, ocean acidification costs, increased drought and wildfire costs, severe weather costs, and increased air pollution costs."

*"A 100% WWS System in each country will **eliminate** such damages."*

When? By 2050... [32]

To suppose that all of these costs will be gone once the 100%WWS Roadmap has been implemented is fallacious. First of all, we've already punched through the sustainable atmospheric 350 ppm carbon dioxide level. In fact, as of August 2016, the level has risen to roughly 405 ppm, and the effects of this transgression will increase as the excess greenhouse gases trap even more heat. Because his roadmap takes 34 years to come to full fruition, we will still be emitting carbon dioxide until 2050 because we've only implemented 4% of his plan so far, and we're already trailing behind on his schedule.

It would be more prudent to say that these costs would be mitigated if the harmful airborne particles (from combustion) have settled down and greenhouse gas levels have stabilized. When could this happen? I don't know, but I'm pretty sure that we will not make it by 2050. Does Jacobson suppose that once we have implemented his roadmap all of these destabilized natural cycles will stop worsening? Are positive feedback loops considered? How is this auxiliary hypothesis proven? These aren't quantified or evidenced. This appears to be little more than a leap of faith in defiance of what is known and understood about climatic processes.

What about mitigating the costs related to the shortage of clean water?

What does the World Health Organisation (WHO) have to say about this?[33]

> ➤ In 2015, 91% of the world's population had access to an improved drinking-water source, compared with 76% in 1990.
> ➤ 2.6 billion people have gained access to an improved drinking-water source since 1990.
> ➤ 4.2 billion people now get water through a piped connection; 2.4 billion access water through other improved sources including public taps, protected wells and boreholes.
> ➤ 663 million people rely on unimproved sources, including 159 million dependent on surface water.
> ➤ Globally, at least 1.8 billion people use a drinking-water source contaminated with faeces.
> ➤ Contaminated water can transmit diseases such diarrhoea, cholera, dysentery, typhoid and polio. Contaminated drinking water is estimated to cause 502,000 diarrhoeal deaths each year.
> ➤ By 2025, half of the world's population will be living in water-stressed areas.
> ➤ In low- and middle-income countries, 38% of health care facilities lack any water source, 19% do not have improved sanitation and 35% lack water and soap for hand washing.

The WHO on sanitation:

> ➤ In 2015, 68% of the world's population had access to improved sanitation facilities including flush toilets and covered latrines, compared with 54% in 1990.
> ➤ Nearly one-third of the current global population has gained access to an improved sanitation facility since 1990, a total of 2.1 billion people. 2.4 billion people still do not have basic sanitation facilities such as toilets or latrines.
> ➤ Of these, 946 million still defecate in the open, for example in street gutters, behind bushes or into open bodies of water.

- The proportion of people practicing open defecation globally has fallen almost by half, from 24% to 13%.
- At least 10% of the world's population is thought to consume food irrigated by wastewater.
- Poor sanitation is linked to transmission of diseases such as cholera, diarrhoea, dysentery, hepatitis A, typhoid, and polio.
- Inadequate sanitation is estimated to cause 280,000 diarrhoeal deaths annually and is a major factor in several neglected tropical diseases, including intestinal worms, schistosomiasis, and trachoma. Poor sanitation also contributes to malnutrition.

What does the UN have to say about water issues?[34]

- Around 700 million people in 43 countries suffer today from water scarcity.
- By 2025, 1.8 billion people will be living in countries or regions with absolute water scarcity, and two-thirds of the world's population could be living under water-stressed conditions.
- With the existing climate change scenario, almost half the world's population will be living in areas of high water stress by 2030, including between 75 million and 250 million people in Africa. In addition, water scarcity in some arid and semi-arid places will displace between 24 million and 700 million people.

As you can see, addressing water scarcity and improving sanitation should be a top priority because it is related to and worsened by man-made climate change.

Additionally, we have to acknowledge that water scarcity is also causing civil unrest and violence: *"The data suggest that the challenges of water conflicts are growing, not shrinking, especially at the sub-national scale. Far better mechanisms and far greater efforts are needed to address these kinds of conflicts."* [35]

Changes in the hydrological cycle precipitated by human activities attribute to man-made climate change, but these simple facts have been omitted. How are we going to counter the fact that half of the world's population will live in water-stressed areas by 2025? We cannot assume that we will be out of the woods once we have eliminated greenhouse gas emissions.

A water-stressed area is an area where water resources are withdrawn faster than they can be replenished. If you continue withdrawing water from your source you will end up depleting the source. To help freshwater resources to remain sustainable, and keep the hydrological cycle intact, we must lower our water usage, reduce water losses in evaporation or leakage, and offset fresh water usage with desalinated sea water. Otherwise, we will reach a point called peak water, after which fresh water resources will literally dry up—or melt irreversibly, in the case of glaciers (which is seemingly inevitable).[36]

With more than two billion people lacking amenities and/or water for sanitation, six hundred million people lacking direct access to potable water, and with fresh water supplies depleting, allotting energy to water treatment, water management, and desalination has to become a priority. However, you can read the entire 100%WWS Roadmap and you will find no mention of these issues. The EIA energy outlook also fails to mention desalination, water management or water treatment.

Here's why: if you consider the amount of energy consumed by mankind, and you put them in a pie-chart, water management would be an unnoticeable sliver—13,451 Mtoe (Million Tons Oil Equivalent) or ~157 TWh in 2013 in total energy versus between 90 and 180 TWh for desalination, or ~1.3% of the total energy supply in 2013.[37] Water management itself is an energy intensive process due to pumping, pressurization, and waste-treatment needs.[38][39] (Not to mention the vast construction costs in terms of energy and materials).

There is always a discrepancy between water withdrawals and water consumption,[40] which means that a lot of fresh water is wasted. If we reduce water usage, we will also reduce water going to waste, and with desalination, we can convert seawater into potable water. Suppose we can alleviate the stress on water resources by desalinating and adding an amount of water to the water network equal to the basic needs of humans.

The WHO states:

*"Based on estimates of requirements of lactating women who engage in moderate physical activity in above-average temperatures, a minimum of 7.5 litres per capita per day will meet the requirements of most people under most conditions. This water needs to be of a quality that represents a tolerable level of risk. However, in an emergency situation, a minimum of 15 litres is required. A higher quantity of about **20 litres per capita per day** should be assured to take care of basic hygiene needs and basic food hygiene. Laundry/bathing might require higher amounts unless carried out at source."* [41]

By 2025, there will be about 8 billion people on Earth, so by that reckoning, we would need to desalinate 40 billion liters of water per day to meet basic water needs, but shouldn't we include water for agriculture as well? According to *"An Overview of the State of the World's Fresh and Marine Waters - 2nd Edition - 2008"*, water withdrawals for agriculture are forecast to rise to roughly 3000km3 per year by 2025.[42][43]

$$3000 \, km^3 = 3,000,000,000,000 \, m^3$$

According to the International Desalination Agency we currently desalinate about 86.8 million cubic meters (M^3) of seawater each day. To achieve this, we need about 0.243 million TWh per day. Which means that the total energy required for current desalination practices is roughly 88.7 TWh. Per year

$$0.243 \, TWh \times 365 = 88.7 \, TWh$$

However, because desalination delivers only 1/3rd of the starting volume[46] as potable water, desalinating 86.8 million M^3 of seawater, we only produce 29 million M^3 of potable water per day.

So how much do we need on a world-scale by 2025?

$$40 \, billion \, Liters = 40,000,000 \, M^3$$

To get 40 billion liters of potable water, we need to desalinate about three times as much seawater: 120 billion Liters, or 120 million M^3. At a theoretical 2.8 KWh/M³, it would require 336 million KWh per day or 0.336 TWh.

$$0.336\ TWh \times 365 = 122.6\ TWh\ per\ year$$

As you can see, desalinating water for humans doesn't take that much energy. In fact, it could be done with the equivalent of 61 GW of electricity from utility-scale PV. However, if we also want to desalinate water for agricultural purposes, the inclination of the curve of requirements quickly steepens. At 1,5 trillion M³ per year (half of the 2025 projected agricultural withdrawal rate).

$$1.5\ trillion\ M^3 \times \frac{2.8 KWh}{M^3} = 4.2\ trillion\ KWh$$

$$4.2\ trillion\ KWh = 4200\ TWh$$

This means that alleviating stress on fresh water sources by desalinating seawater, will require roughly 4300 ~ 4500 TWh per year. (I am omitting the energy need for pumping, storage, and waste-water treatment.) Under normal circumstances, it is economically unattractive to pump water over vast distances. However, if you can alleviate stress on fresh water sources, I think it is well worth it.

Desalination, increased water management, and water treatment are essential while we improve our water infrastructure and become more efficient with water for our agricultural, industrial, and domestic processes. However, we must include the energy needed to get started. If we fail to address the water problems, the 100% roadmap will be for naught, and many will die from the effects of water scarcity and the lack of sanitation. Unfortunately, the increasing stress on water supplies has doesn't appear in the 100%WWS roadmap.

In conclusion, the 100%WWS roadmap fails to provide the energy required to get this world-spanning water management effort going. What we see again is that this proposed model is hopelessly divorced from the challenges we must face and manage in the actual, real world. Proposing solutions for a world that doesn't and will not exist is not just an exercise in futility, but a dangerous distraction.

Material requirements

Unfortunately, our power plants and energy "capture devices" like wind turbines and solar panels that don't really generate energy—require a lot of resources and/or feedstock to operate. Each technology has its own footprint, which is determined by a number of factors. Consider for instance the capacity and capacity factor. Capacity factor is determined by the amount of mean time that a generator delivers its maximum power.

How many Kg of materials would we have to invest to get a life cycle worth of energy from 1GW worth of SunPower E20 435Watt panels? A PV panel has an operational life of about 20 years, and 1GW worth of PV panels normally would deliver roughly 2,507 GWh in a year and 62,675 GWh during 25 years. How much does such a PV panel weigh? According to SunPower's factsheet 25.4 Kg.[47] How many kg of finished end-product materials are required per GW?

$$435\ W = 0.000000435\ GW$$

$$\frac{1\ GW}{0.000000435\ GW} = 2{,}298{,}850\ panels$$

$$2{,}298{,}850\ panels \times 25.4\ kg\ per\ Panel = 57{,}471{,}264\ kg$$

With this information, we can calculate the minimum material requirements per GWh of energy produced over the lifetime of the panels.

$$\frac{57{,}471{,}264\ kg}{62{,}675\ GWh} = \frac{916\ kg}{GWh}$$

However, this figure is incomplete because solar panels must be mounted, which requires more materials. Additionally, utility-scale PV plants have even larger material requirements because they are built on the ground, not on existing roofs. For instance, consider the amount of concrete required, the amount of steel and aluminum for the mounts and the copper required for the wiring. If we consider table 10.4 of the EIA's 2015 Quadrennial Technology Review, solar requires 16,447kg[48] of materials per GWh of generated electricity over the 20 year lifespan of the PV technology.

I use the figures from a document called "Life cycle assessment of utility-scale CdTe PV balance of systems". According to this document, the additional material requirements for utility scale PV plants are 16 kg per square meter of panel. The SunPower E20 430Watt panel has a surface area of about 2m². By this reckoning the total surface area of a 1GW PV plant would be 4,597,700 m², and the additional mounting and cabling requirements would be 73,563,200 kg.[49]

$$\frac{131,034,464\ kg}{62,675\ GWh} = \frac{2090\ kg}{GWh}\ for\ Utility\ Scale\ PV$$

Rooftop PV systems require less material because the mounting requirements are much simpler. Roof mount figures vary per system used. For instance, if the roof has the correct inclination and is facing the equator (south-facing for north of the equator and north-facing for south of the equator) very limited materials are required. However, if you have a flat roof, you'll need to raise the panels to the proper inclination. Adding a solar tracking system further raises material costs. However, in this case. let's subtract the foundational part of the material mounting requirements for utility-scale PV, which will give us a total of 1.5 kg of mounting materials required per square meter. By this reckoning the total surface area of 1 GW of rooftop PV would be 4.6 million m², and the additional mounting and cabling requirements would be 6.9 million kg (remember, we had an initial material requirement of ~57,47 million kg for the panels).

$$\frac{64,367,814\ kg}{62,675\ GWh} = \frac{1027\ kg}{GWh}\ for\ rooftop\ PV$$

So what about wind? According to the 2011 report "Wind Energy in the US and Materials Required for the land-based Wind Turbine industry" by the US Geological Survey, next generation wind turbines will require 540,000 kg of materials per MW of capacity for an onshore wind turbine.

A 2010 paper called "The carbon dioxide footprint of offshore wind" roughly confirms this figure. See page 8, table 2-2. The same table provides the following figures for offshore wind: rotor 38 tons, nacelle 64 tons, tower 140 tons, foundation 203 tons. If we add these up (and keep in mind that this is a Vesta V80 2MW unit), we get a total of 222,000 kg of materials per MW of capacity for an offshore wind turbine.[50][51]

1GW worth of wind turbines produces about 2849 GWh in a year and 71,223 GWh during their 25-year operational lifespan.

$$\frac{540{,}000{,}000\ kg}{71{,}223\ GWh} = \frac{7582\ kg}{GWh} \text{ for onshore wind}$$

$$\frac{222{,}000{,}000\ kg}{71{,}223\ GWh} = \frac{3116\ kg}{GWh} \text{ for offshore wind}$$

Technology	Total yield in TWh per year	Materials in Ton/TWh	Required Materials in Metric Tons
Onshore Wind	18,296	7582	139 million
Offshore Wind	10,860	3116	34 million
Res. Roof PV	9870	1027	10 million
Com/Gov Roof PV	10,728	1027	11 million
Utility Solar PV	61,253	2090	128 million
Average		2968	
Total	111 thousand		322 million

Later in this book, we will contrast these numbers with other technologies. This exercise is intended to demonstrate that material costs are tied to all energy technologies. This is the basis for the *efficiency argument*. We need to do more with less, not less with more. By looking at the material footprint per unit of energy produced over the lifespan of the technology we can get a complete picture, rather than working from a material footprint per capacity viewpoint.

It is also important to realize that cumulative upkeep has been omitted from this picture. We'd have to replace or rebuild these units after 20 to 25 years or whenever there are problems. Therefore, the presented figures are low estimates.[52]

Finally, keep in mind that the materials required for these technologies have to be extracted from the Earth. This exercise will give you a scope of the volumes involved and how often these materials will need to be provided.

Over committing to solar, wind, hydrogen, and batteries is the most fundamental flaw in the 100%WWS roadmap because it will greatly stress natural resources, especially when the required growth curve is as steep as projected by the roadmap. The roadmap fails to address lifecycle issues that include costs of installation, maintenance, decommissioning, and recycling emissions; it fails to address the fact that one cannot recycle 100% of all the materials used, and that getting closer to 100 will require more energy and specialized chemicals that we don't have. In fact, the 100%WWS Roadmap actually calls for a decline in energy availability, not an expansion. Therefore, the circular principle is dead in the water before the 100%WWS Roadmap can be deployed.

"According to the U.S. Department of Energy's Critical Materials Strategy report, there are several critical materials that are crucial to the development of solar technologies. Although used in small amounts, rare earth minerals, and other critical minerals are an essential element to the development and manufacturing of solar technologies. Due to this factor, and the current lack of viable alternatives, rare earth minerals, and other critical minerals are essential to the production of solar technologies."

http://www.seia.org/policy/manufacturing-trade/critical-materials-rare-earths

Which materials and elements are most needed in the 100%WWS Roadmap?

Critical Materials Found in Clean Technologies

Technology	Component	Material
Wind	Generators	Neodymium
		Dysprosium
Vehicles	Motors	Neodymium
		Dysprosium
	Li-ion Batteries (PHEVs and EVs)	Lithium
		Cobalt
	NiMH Batteries (HEVs)	Rare Earths: Cerium, Lanthanum, Neodymium, Praseodymium
		Cobalt
PV Cells	Thin Film PV Panels General*	Tellurium
		Gallium
		Germanium
		Indium
		Selenium
		Silver
		Cadmium**
	CIGS Thin Films	Indium
		Gallium
	CdTe Thin Films	Tellurium
Lighting (Solid State and Fluorescent)	Phosphors	Rare Earths: Yttrium, Cerium, Lanthanum, Europium, Terbium
Fuel Cells*	Catalysts and Separators	Platinum, Palladium and other Platinum Group Metals, Yttrium

Sources: Table data extracted from Bauer, 2011 (20) and expanded upon with data from other sources per asterisks. *APS/MRS, 2011 (2). **Lifton, 2011 (10)

Source & Image Credit: http://resnick.caltech.edu/docs/R_Critical.pdf

All of these materials need to be extracted from the earth, processed and transported. It is also important to note that mining, besides being hazardous, poses a danger to the surroundings and contributes to climate change. Some of these materials are in very high demand, but availability is low and is projected to remain low.

Here are some excerpts and conclusions from different relevant scientific studies on the availability of materials essential to the 100%WWS Roadmap.

"One need only look to the growth in demand for copper (Cu) for an example of the importance of long-term thinking when it comes to materials strategy. China now uses 40% of the world's Cu, compared to just 6% in 2000, an astounding increase over 10 years. If their current growth demand continues, they will required the equivalent of the worldwide 2010 Cu production by 2018. While ample resources exist, it may not be possible to increase Cu production to meet the worldwide demand (see Figure 4). This discrepancy could hamper the expansion of renewable energy activities throughout the world, as well as limit the growth of China's economy.

The increased volatility in the supply of copper will also likely lead to an even more jumbled picture for several technology metals that are co-produced with it, including tellurium, selenium, and rhenium. These materials are already deemed critical due to the unstable nature of this coproduction, which could be even more unstable if Cu production is squeezed." [53]

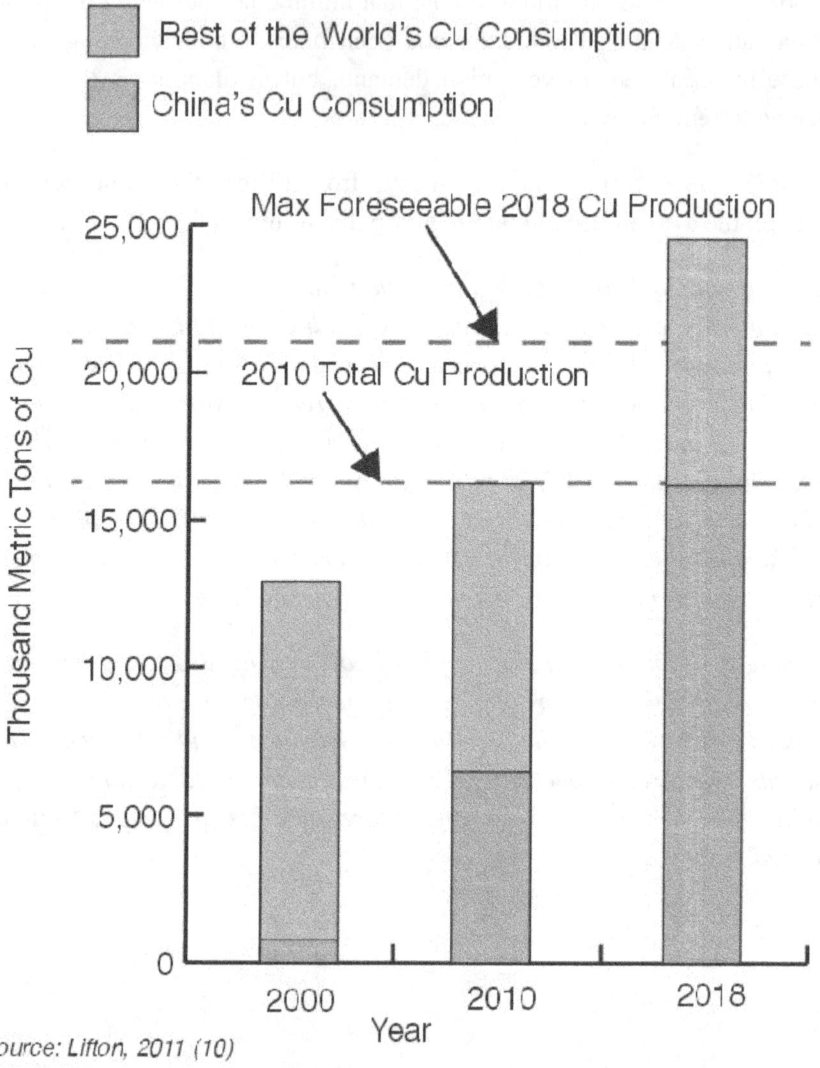

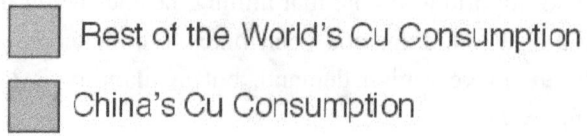

Source: Lifton, 2011 (10)
Source & Image Credit: http://resnick.caltech.edu/docs/R_Critical.pdf

According to the January, 2016 Copper mineral summary by the U.S. Geological Survey National Minerals Information Center, we're not even close to reaching 21~22,000 thousand metric tons of Cu production. Current production levels are somewhere between the 18,700 and 19,000 thousand metric tons. And <u>worldwide</u> production grew by 200 thousand metric tons in 2014. The 100%WWS Roadmap calls for a rapid expansion of the existing power grids, which will increase demand for copper as will any increase in PV production.

The Copper Development Association estimates that each renewable system (wind and solar) will require 5.5 metric tons of copper.[54][55][56]

Wind and PV additions will eventually require about 2,561 GW of additional power per year. By this reckoning, wind and solar would require this amount of copper:

$$2,561,000 \times 5.5 = 14,085,500 \text{ metric tons of Cu per year}$$

Or

$$\frac{14,085,500 \text{ metric ton}}{1000} = 14,085 \text{ thousand metric tons}$$

14,085 thousand metric tons is slightly more than the total Cu production in 2010. Cu projections vary, but none suggest that we have the copper required to facilitate the growth required by the 100%WWS Roadmap.

Let's contrast the required amount of copper with what was used last year. Renewable additions where roughly 63 GW for Wind and 50 GW for solar PV according to the Renewables 2016 Global Status Report by REN21.

A grand total of 113 GW of wind and solar was added in 2015.

$$113,000 \text{ } MW \times 5.5 \text{ } tons = 621,500 \text{ metric tons of Cu in 2015}$$

The total production of Cu was 18,700,000 metric tons.

$$\frac{621,500 \text{ metric tons}}{18,700,000 \text{ metric tons}} \times 100 = 3.3\%$$

3.3% of all copper produced in 2015 was used to build wind turbines and solar panels and to connect them to the grid.

What about electrical motors required for both BEV's and FCEV's which also require copper? Business Insider provides a useful graphic. [57]

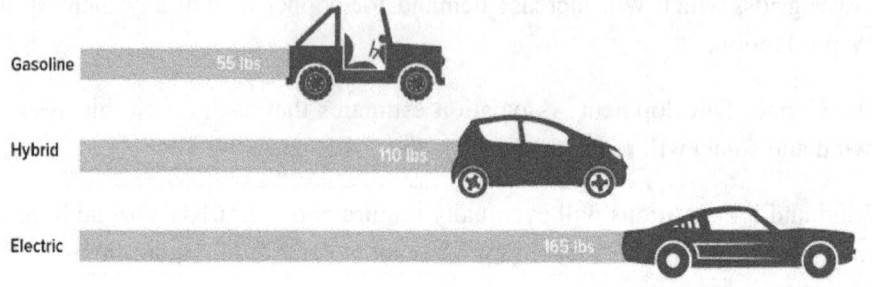

Source: Visual Capitalist, U.S. Global Investors

Confirmation of these figures can be found in: *"How much copper does that electric car need"*[58]—*"between 150 and 180 pounds"*.

Suppose we can recycle all the copper from gasoline vehicles, and re-use it to build electric vehicles and fuel-cell vehicles. And suppose that the amount of copper used in BEV's and FCEV's is roughly the same. We could make do with 110 pounds of copper per vehicle. 110 pounds roughly equals 50 kilograms. We'd need one billion of these if we would want to replace all cars with EV's, and we'd be building these for 34 years, which means that we would be building 30 million EV's a year, using 1471 thousand metric tons of copper each year on top of the 14 000 thousand metric tons of copper required to realize the 100%WWS Roadmap.

Let's tally these figures:

2015 copper production	18,700
2015 copper usage in RE	625
2020 copper usage PV + Wind	14,000
2020 copper usage EV's	1500
Total annual required per 2020 (very low estimate)	34,200 thousand metric tons

Note that we've omitted the need for additional electrical buses, trains, subway cars, boats, and airplanes. It is estimated that electric trolleys, buses, and subway cars use an average of 2300 pounds or 1050 kilograms per unit[59].

Let's consider this chart from the 2014 World Copper Factbook.

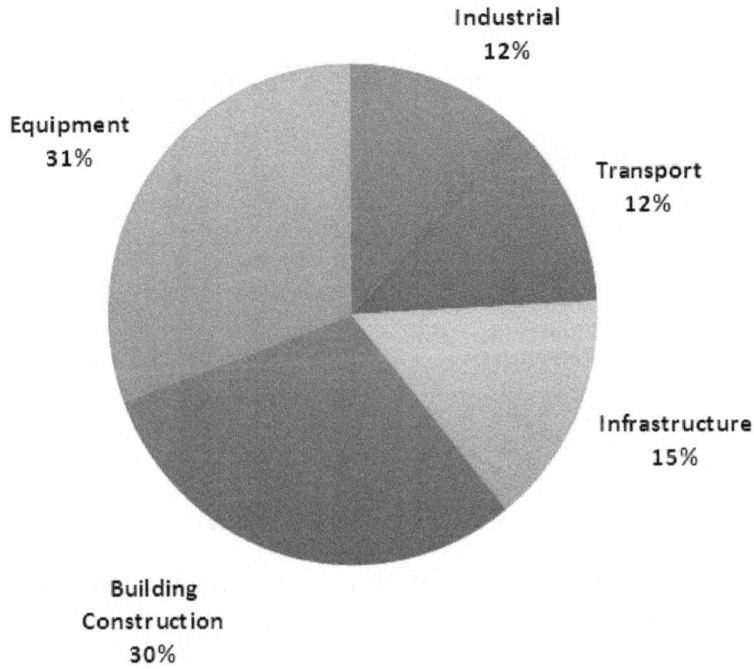

Source: http://copperalliance.org/wordpress/wp-content/uploads/2012/01/ICSG-Factbook-2014.pdf

Copper has many uses: wiring; in semiconductors as circuitry; in piping and plumbing; as a building material; in electric motors; we even use it as a hull material for boats, but renewables have only had a marginal share of all available copper. However, that will change if we choose to implement the 100%WWS Roadmap. Copper use for wind and solar is currently only 3.3%, but if we adopt the roadmap plan it could rise to 75~80%. The limitations of copper production might just be the straw that will break the camel's back.

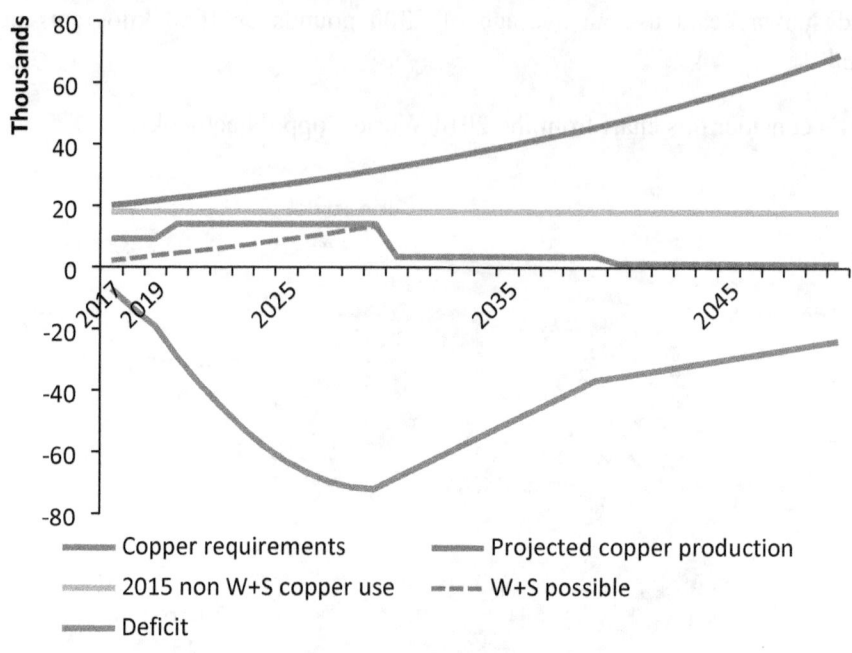

Figures are in Thousand Metric Tons - this is the 100%WWS Roadmap requirement i.e. 118,000 TWh.

I am cheating a little here. I've used data that will be presented later on in this book to plot a graph to show the possible growth of copper production (**brown**) based on an annual growth rate of 3.8% (which is as stated by the USGS); You can see how much copper is required per year for all the wind turbines and solar panels required in the 100%WWS Roadmap (**blue**); You can see the total copper usage in 2015 (**green**); you can also see how much wind and solar additions are possible per year (**purple striped**) and within this 3.8% annual growth scenario for worldwide copper production; and you can see the deficit (**red**), which is a function of the cumulative required and possible build rate. As you can see, by 2050 there is still a clear deficit, which means that, unless we can somehow boost copper production, the 100%WWS Roadmap is impossible to implement.

It is fair to build in some room for uncertainty. Because we are unsure about the amount of energy that will be required by the 2050s, we could extrapolate five possible end-values, these being: 100,000 150,000 200,000 and 250,000 TWh/yr, but first, we have to determine the required growth rate for each scenario.

$$\text{Growth Rate} = \left(\frac{\text{End Generation Required}}{\text{Start Generation capabilities}}\right)^{1/N} - 1$$

Where N is the number of years, which is 34. This gives us the following graph.

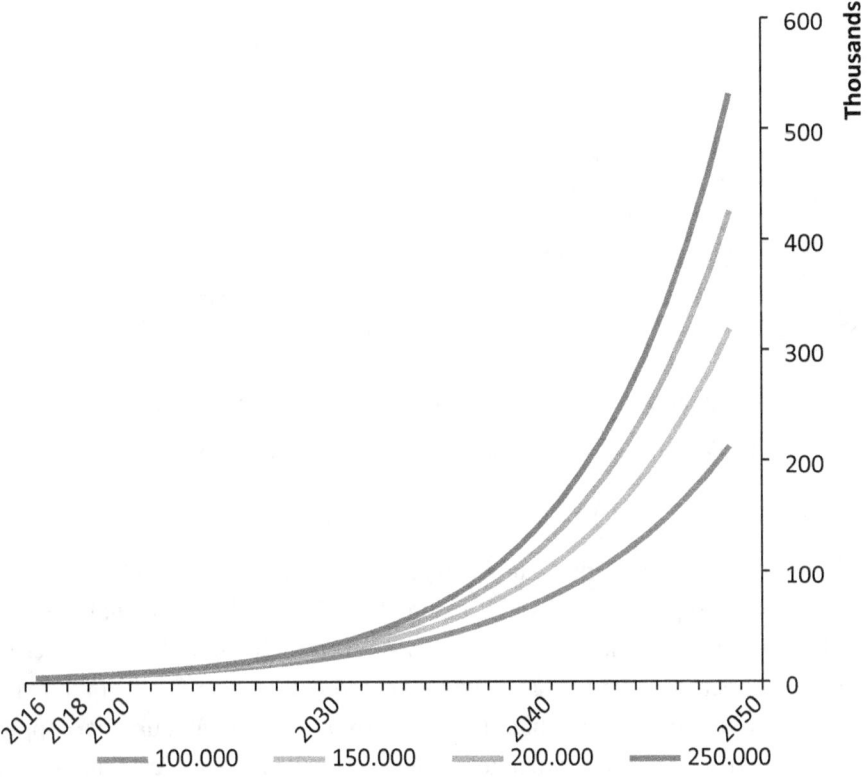

Figures are in Thousand Metric Tons

These projections are based on growth rates ranging between 12.6% and 16.4%. However, if we would implement the roadmap based on these growth rates, we'd be decarbonizing far too slow. Let's implement the 100%WWS Roadmap's curve for each end target, the end figures would remain the same, however, the inclination would be entirely different.

it will be the same as the inclination of the figure shared by the solutions project i.e. 20%@ 2020, 50%@2025, 80%@2030, 95%@2040, 100%@2050 -- source: http://thesolutionsproject.org/resource/transition-chart-to-100-clean-renewable-energy

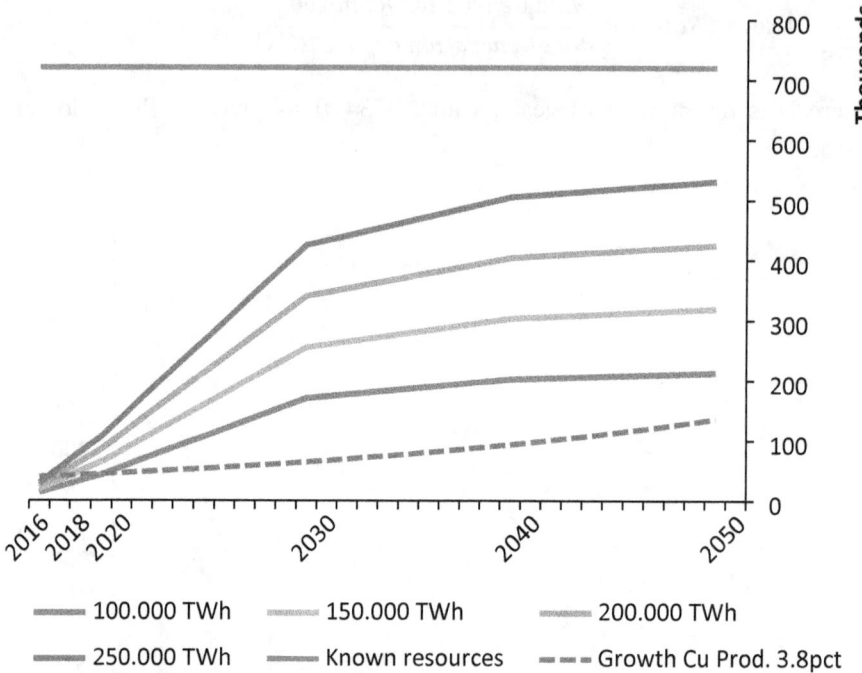

Figures are in Thousand Metric Tons

It is important to note that known reserves are roughly 720,000 thousand metric tons. Suppose we could reach 100,000 TWh of annual energy production on Wind and Solar, and it would require 200,000 thousand metric tons of copper, we would have extracted more than one-fourth of all the copper known to be available. It is not so much the amount of copper required that makes this challenge so big, it is about the increase in copper production <u>growth</u> required, that makes it hard to achieve. As you can see, copper production trails behind in every scenario. Also, we're mining deposits that have lower copper densities, so increasingly more energy has to be invested per unit of copper.

The USGS states the following regarding substitutes: *"Aluminum substitutes for copper in power cable, electrical equipment, automobile radiators, and cooling and refrigeration tube; titanium and steel are used in heat exchangers; optical fiber substitutes for copper in telecommunications applications; and plastics substitute for copper in water pipe, drain pipe, and plumbing fixtures."*

The most important alternatives aren't mentioned: Base plates and contact points for PV and wiring for engines and generators. Besides, aluminium is less efficient than copper in a lot of cases. If we substitute aluminium for copper, we would only be shifting the burden.

Even if we keep production as high as possible, we wouldn't be able to build all the wind turbines and solar panels required before 2050 due to the deficit in copper. Also note that we are already behind 1500 GW of annual required wind and solar additions (which is about 15 times more than we currently add).

Shortages of rare earths already plague the renewable industry. Research institutes confirm that there's a tremendous strain on resources already and foresee a shortage of materials in the not-so-distant future. I doubt that we can sustain the raw material production levels that are needed to facilitate this rapid growth in wind and solar as required in the 100%WWS Roadmap. And research and development on raw material recovery is lagging, so we may conclude that the 100%WWS Roadmap is already facing serious startup problems.

Consider this 2015 article: *Heavy rare earths, permanent magnets, and renewable energies: An imminent crisis.*[60]

"This article sounds the alarm that a significant build-out of efficient lighting and renewable energy technologies may be endangered by shortages of rare earths and rare earth permanent magnets. At the moment, China is the predominant supplier of both and its recent rare earth industrial policies combined with its own growing demand for rare earths have caused widespread concern. To diversify supplies, new mining—outside of China—is needed. But what many observers of the "rare earth problem" overlook is that China also dominates in the processing of rare earths, particularly the less abundant heavy rare earths, and the supply chains for permanent magnets. Heavy rare earths and permanent magnets are critical for many renewable energy technologies, and it will require decades to develop new non-Chinese deposits, processing capacity, and supply chains. This article clarifies several misconceptions, evaluates frequently proposed solutions, and urges policymakers outside of China to undertake measures to avert a

crisis, such as greater support for research and development and for the cultivation of intellectual capital."

And from the same document:

"This article seeks to broaden awareness of a potential threat to a significant build-out of efficient lighting and renewable energy technologies: the almost monopolistic dominance of one country, China, over rare earth processing, particularly for heavy rare earths, and permanent magnet production. The policy implications of this dual dominance are significant. Innovation, which is clearly needed to catch up to China, is fostered by supporting 'niche-level' activities (Geels, 2010) such as research and development, laboratories and information exchange. Policy makers outside of China should clearly step up their support. Moreover, not only does China have a substantial head start regarding its heavy rare earth processing and permanent magnet industries, it has other 'assets' that are lacking elsewhere: technological know-how and intellectual capital. Technological know-how can be acquired, but intellectual capital—the engineers and scientists specialized in these fields—must be cultivated and developed."

Neodymium is required to make the permanent magnets used in windmills. According to a paper called *Peak Neodymium-Material Constraints for Future Wind Power Development*[61] *"there is a high degree of certainty that there are serious limitations on wind power growth, thanks to the limited availability of Neodymium."*

"the best permanent magnet is neodymium-iron-boron (Nd2Fe14B) based magnets, which also contain certain amount of praseodymium, and smaller quantity of dysprosium and terbium (Schüler et al., 2011). According to the study carried out by Schüler's team (2011), neodymium based magnets has the advantage of high energy product that can reach 400 kJ/m3 or more (Cullity and Graham, 2008, p. 491), being about 2.5 times higher than samarium cobalt magnets and 7-12 times stronger than aluminium iron magnets. Meanwhile, additional rare earth metals such as dysprosium and terbium (Goonan, 2011) are added to the magnets in order to overcome the corrosion problems as well as limitation of operation temperature (Müller et al., 2001)."

"Consequently, the growing popularity of permanent magnet generator leads to increasing demand of Nd. Based on Emsley's (2011) statement, wind turbines armed with permanent magnets require 0.7-1 ton of neodymium alloy for every megawatt (MW) of capacity. And a single Scanwind 3500 DL wind turbine with a 3.5 MW capacity, produced by a Finnish company called The Switch needs more than 2 tons (equal to approximately 0.6t/MW produced) of neodymium-based (Nd-Fe-B)permanent magnet material for manufacturing (Hatch, 2009). In order to achieve enough wind power based electricity supply for global from Wind, Water and Sunlight (WWS) system, an increase by a factor of more than 5 in annual neodymium world production would be needed, which is quite impossible to be realized for a long time even with new extraction along with recycling measures (Jacobson and Delucchi, 2011). Additionally, more constraints from political power and incentives resulting from environmental concerns will limit the expansion of supply in the future (Lifton, 2009). The U.S Department of Energy (DOE, 2011) conducted a criticality assessment of rare- earth metals and pointed out supply challenges for dysprosium, neodymium, terbium, and yttrium in terms of clean energy technologies. Rare-earths permanent magnets benefits larger turbines and slower turbine speeds with direct-driven arrangement. Both of these designs are regarded as main trends of wind power development."

"Meanwhile, the conventional wind power plant can also cause growth in demand of neodymium apart from the direct-driven ones. Since permanent magnets are also capable of reducing weight and cost of conventional wind turbine construction. An example of neodymium usage is that it is able to reduce an amount of weight of 10 tons of steel in the V112–3.0 MW tower (Davidsson et al.,2012). Different from direct-driven gearless wind turbines in which the neodymium in the form of permanent magnet is irreplaceable, conventional turbines require much less neodymium or even can be free from it. But more and more conventional designs are implementing permanent magnets to increase the efficiency and reduce the weight. Thus, neodymium utilization in conventional wind turbines should not be neglected."

"In fact, the criticality of neodymium along with other rare earth metals used in wind turbines manufacture is less mentioned in current discussion or assessment of wind power plant constructions. More generally discussed *issues are environmental and social impacts caused by wind power plant construction. These impacts mainly include sound propagation (Pedersen and Halmstad, 2003), health disturbance (Colby et al., 2009), threats to wildlife (Kuvlesky et al., 2007), increasing demand in land, and so forth (Wizelius, 2006, p.127- 205)."*

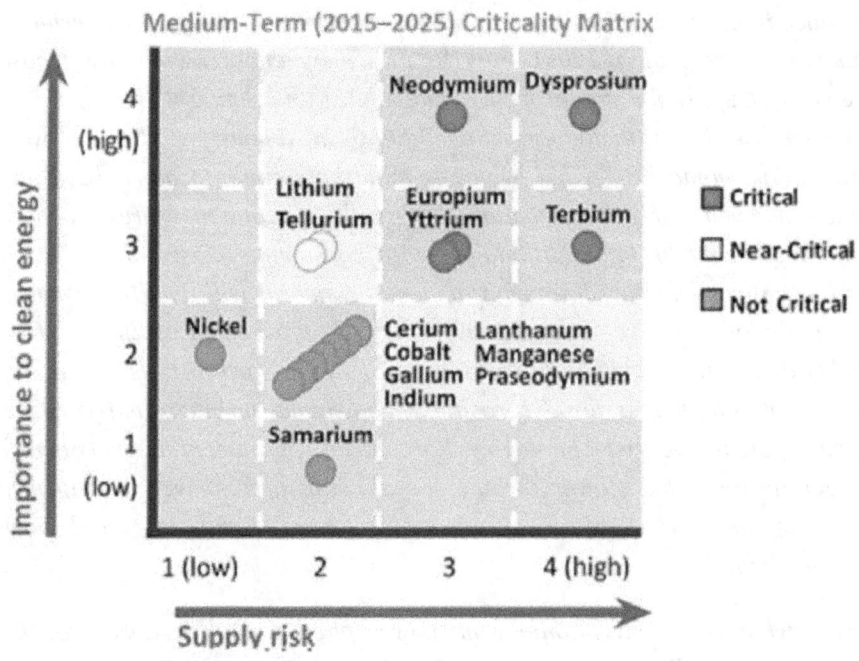

Short-Term (2011-2015) and Medium-Term (2015-2025) REE Criticality Matrix (DOE, 2011)

As you can see, the build-out of wind and solar will be difficult as the demand for special materials like neodymium, but also in ubiquitous materials such as copper or silicon, increase. There are limits to what we can extract from the earth, and how fast we can get the materials required.

"Rare earth elements (REE) such as scandium, yttrium and 15 other minerals are sought after for their unique technology applications, but their scarcity means those countries lucky enough to contain them hold significant sway over global supply."[62]

"Nevertheless, we are mining poorer and poorer ores all the time, and it takes more and more energy to extract the same amount of metal."[63]

If we want to do an all-out build-off bonanza as required in the 100%WWS Roadmap, we will be faced with massive shortages so long as mining capabilities trail behind the demand. Lake Baotou in China, which is a dystopian place, is one of the shady sides of rare earth mining. Once the rare earth materials have been mined, you have to separate them from the ore, which is done by a host of different chemical processes with hazardous tailings and other valuable elements going to waste.

"Processing rare earths is a dirty business. Their ore is often laced with radioactive materials such as thorium, and separating the wheat from the chaff requires huge amounts of carcinogenic toxins – sulphates, ammonia and hydrochloric acid. Processing one ton of rare earths produces 2000 tons of toxic waste; Baotou's rare earths enterprises produce 10m tons of wastewater per year. They're pumped into tailings dams, like the one by Wang's village, 12km west of the city centre."[63]

Note that thorium is only mildly radioactive, and hardly hazardous at all. But for the sake of sensationalism, it is added to the list of evil elements that spread through lake Baotou.

"The environmental risks and impacts associated with metals extraction, processing and refining are many, and a rising demand for metals will inevitably have major environmental and social impacts, including significant contribution to climate change. Environmental risks are expected to grow if new mines with sub-standard environmental protection are opened in a hurry to meet demand, and as extreme weather events become more frequent."[64]

The people who work in the mines and live near the refineries pay the price for our technologies. So let's go for the technologies that are most efficient in terms of materials used. As long as we fail to clean up our mining and purification practices, *"clean energy"* will be an oxymoron.

However, we already have a safe, efficient, environmentally friendly technology for generating electricity—nuclear energy, especially when produced by modern plants that can "burn" our stored nuclear waste as fuel. More on this later.

Run the numbers

We need a paradigm shift in energy generation. We have to move away from coal, gas, and oil, and replace them with alternatives that provide as much or more energy to support our growing, industrious civilization. However, in making energy choices, remember that nothing matters more than the numbers.

To engage in geoengineering, reverse the carbon debt through permanent sequestration, remedy ocean acidification by capturing CO_2, and provide more potable water through desalination, we will need to add ~12,000 TWh.

We add this figure to prediction A: the 100%WWS Roadmap, ~130,000 TWh; and prediction B: the EIA per 2040 expected annual demand, ~240,000 TWh, which might not seem like a lot. But because our current electricity generation is about 25,000 TWh, adding 12,000 TWh to this figure is impressive.

Let's project current production levels until 2050 to see if they will suffice. Following that, we'll determine how much production will be needed to reach the 100%WWS figure and the EIA figure, plus the power required for desalination, sequestration, and de-acidification.

First consider one of the possible growth scenarios distilled from several sources, amongst which the EIA, by Jacobson and others.

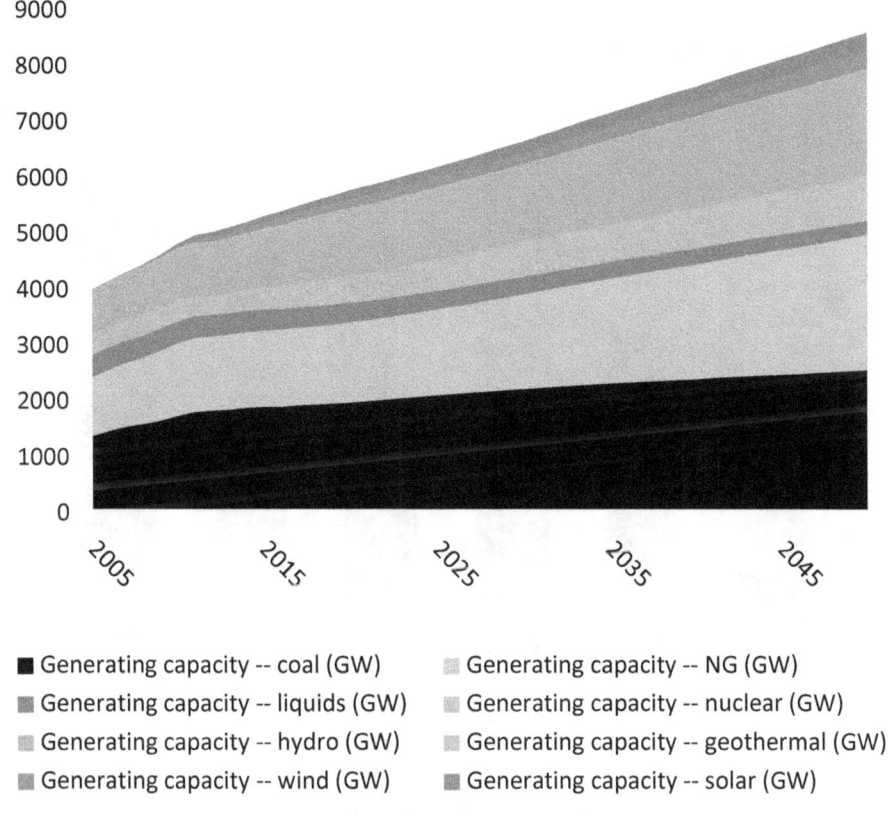

- ■ Generating capacity -- coal (GW)
- ■ Generating capacity -- liquids (GW)
- ■ Generating capacity -- hydro (GW)
- ■ Generating capacity -- wind (GW)
- ■ Generating capacity -- NG (GW)
- ■ Generating capacity -- nuclear (GW)
- ■ Generating capacity -- geothermal (GW)
- ■ Generating capacity -- solar (GW)

Total capacity installed by 2050: 9284 GW[65]

Note that only one technology is expected to decline: Liquids—which means electricity from "oil"—mostly diesel.

In this scenario, our energy mix would deliver 44,000 TWh by 2050, but this is not a fully electrified scenario because we're short tens of thousands of Terawatt hours. In this case, we would still be burning a lot of fossil hydrocarbons for electricity and thermal processes. By EIA's 2050s prediction, we would still be using the equivalent of roughly 190,000 TWh for thermal, non-electrical, processes.

This is what the EIA predicts.[66]

Figure 5-3. World net electricity generation by fuel, 2012–40
trillion kilowatthours

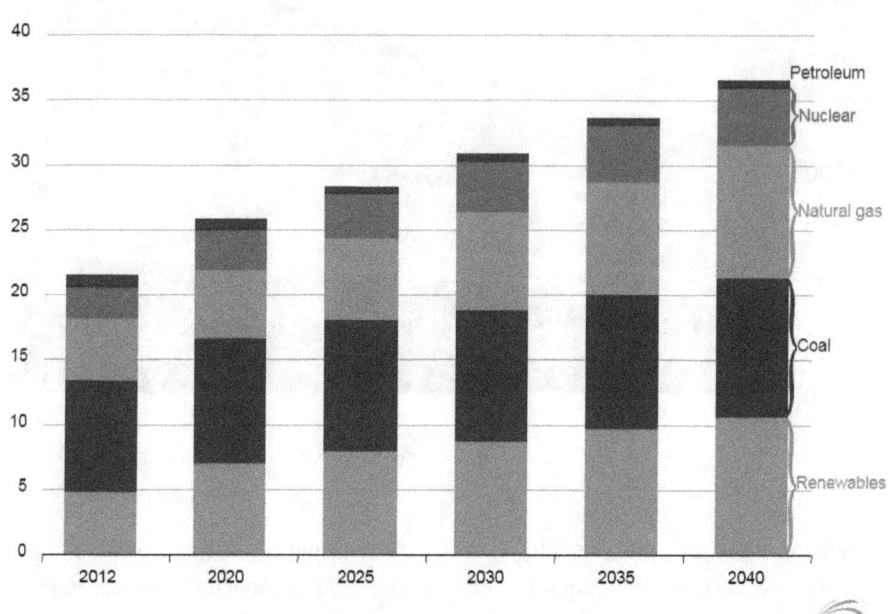

If these projections are accurate, renewables will hardly displace coal and utterly fail to displace natural gas because of population growth and the need for more energy. Also, note that this EIA forecast only reaches 36~37,000 TWh of electricity generation. Which is a small portion of our total energy consumption.

With my preference for the xWh [energy] metric over the xW [capacity] metric in mind, let's ask how much energy can we expect this mix to generate, based on the capacity factors previously mentioned?

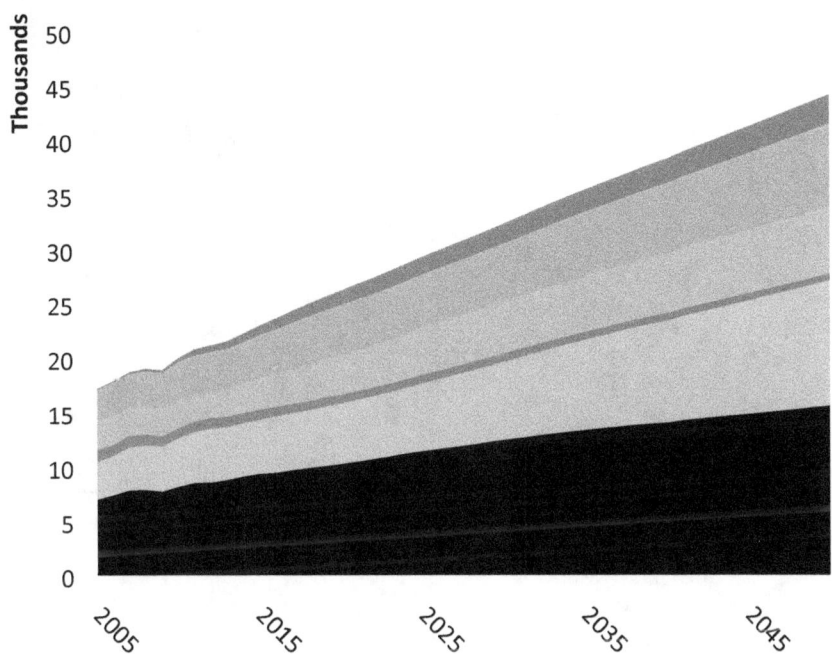

Figures in TWh

Here, we encounter a serious problem. The EIA, among other agencies, expects electricity generation to be no more than 44,000 TWh, which gives us a discrepancy of roughly 74,000 TWh with the roadmap's fully electrified hypothesis. Also, note that roughly one-third of this EIA energy prediction is supposed to be non-carbon emitting. However, this means that we would still be burning a considerate amount of fossil fuels by 2050, and it is unclear what the climate will be like by then, and what the repercussions for the biosphere and mankind would be. Is this a gamble we are willing to take? From this perspective, I agree with Jacobson that it is absolutely necessary to eliminate fossil fuels as quickly as possible. The proposed alternative pathway however, is different and probably more effective.

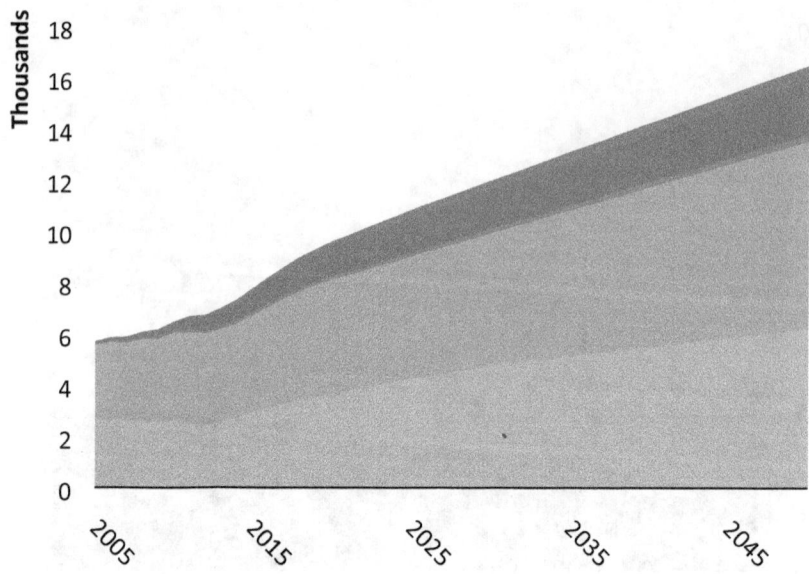

Figures in TWh

This graph is the same as the previous—without carbon fuels like coal, gas, and liquids. Here, the non-carbon technologies barely reach 16,000 TWh. This is a 90,000+ TWh discrepancy with the Roadmap. Wind and Solar are only one fifth portion of the aforementioned energy mix. In fact, it is more than 100,000 TWh short of the roadmap, without the thermal storage. To be honest, this scenario is too negative in regard to the projected rate of growth of wind and solar power. However, these technologies face several problems which may hamper future implementation rates.

Also note that wind and solar have to play a role in reaching a future free from fossil fuels because we are still moving forward scientifically, technologically and culturally and this will require massive amounts of energy. This doesn't mean that the 100%WWS roadmap would stop scientific process, but it does hamper progress when energy becomes more expensive and less available.

The following chart illustrates what the growth curve of the 100%WWS Roadmap looks like. It is not as fancy as the one available at the solutions project website, but it provides a better understanding of the challenge ahead.

Note that the values on the vertical axis are in TWh, and that I've grouped residential, commercial, and government roof PV together. From analyzing this graph it becomes clear that tidal, wave, geothermal and hydropower are a mere sliver of the total energy mix that is proposed in the roadmap. 96.3% of this mix is wind (24.6%) and solar power (71.7%).

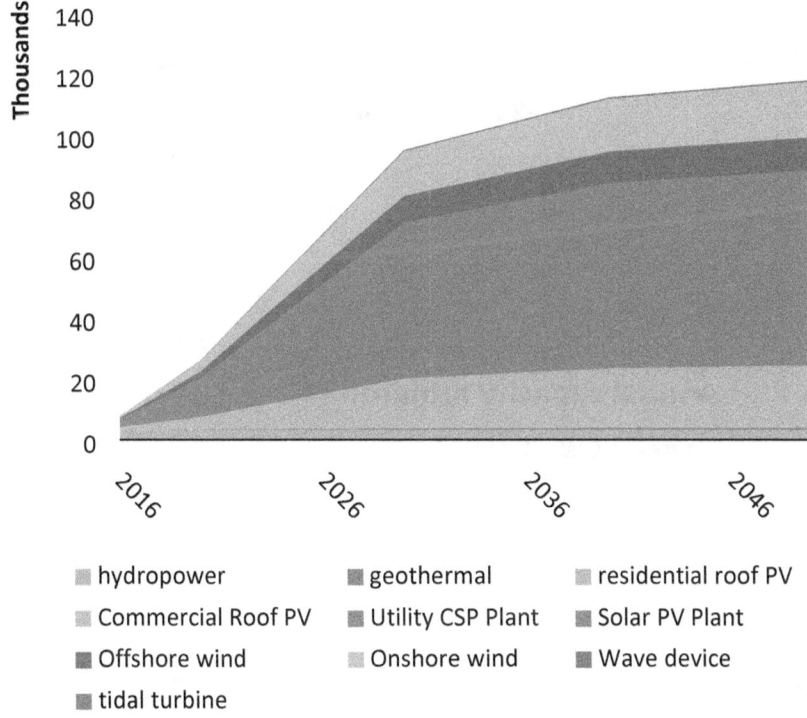

Figures in TWh

We are slowly arriving at the point where the unlikelihood of this roadmap being feasible is coming to light, and we are now going to contrast required growth with actual growth.

Annual generation capabilities required

Figures in TWh growth per year

Period	Geothermal	Solar	Wind	CSP
2016-2020	23.9	3238	1153	123
2020-2035	36.1	4889	1741	186
2025-2030	26.2	4889	1741	186
2030-2040	9.0	1222	435	46
2040-2050	3.0	407	145	15

Annual capacity required

Figures in GW capacity added per year

Period	Geothermal	Solar	Wind	CSP
2016-2020	3.8	1292	405	62
2020-2035	5.7	1950	611	94
2025-2030	4.2	1950	611	94
2030-2040	1.4	488	153	23
2040-2050	0.5	163	51	8

As you can see, the 100%WWS Roadmap mainly depends on a solar capacity growth rate of 1292 to 1950 Gigawatt per year, and a wind capacity growth rate of 405 to 611 Gigawatt per year. To make sense of these numbers I will present the growth rate of these technologies in 2015 according to the *"Renewables 2016 Global Status Report"* by REN21.[67]

Highest actual capacity added

Figures in GW

Year	Geothermal	Solar	Wind	CSP
2015	0.6	50	63	0.5

Fantasizing about a world that is powered by solar and wind might land you at the table of incredibly wealthy superstars, but it doesn't provide a realistic model. The roadmap's start capacity growth for solar is 25 times higher than we have managed to realize, and wind's capacity growth is 6 times higher. The required additions of concentrated solar power, however, eclipse past additions by 120 times. If the challenge is 240,000 TWh instead of 118,000 TWh, double all figures if you want to make it before 2050, which we must do.

The 100%WWS Roadmap also makes an overly optimistic appraisal of the lifetimes of PV and Wind technologies. Although the EIA states 20 years for both wind energy and solar PV, a lifetime of 35 years for wind and 30 years for PV plants are claimed in the roadmap. We have yet to see the first wind turbine facility to reach 25 years of operation. In fact, the Cowley Ridge wind facility in Canada, which is the oldest in the world, will be taken offline well within the 25 year lifetime – and it won't be repurposed.

Now let's see what happens if we stop production of wind turbines and PV panels and other technologies at 2050. If we assume that by 2050 max lifetime of units produced in 2016 is reached. (I could have used percentages to plot this graph, but I want to have some sharp identifiers.)

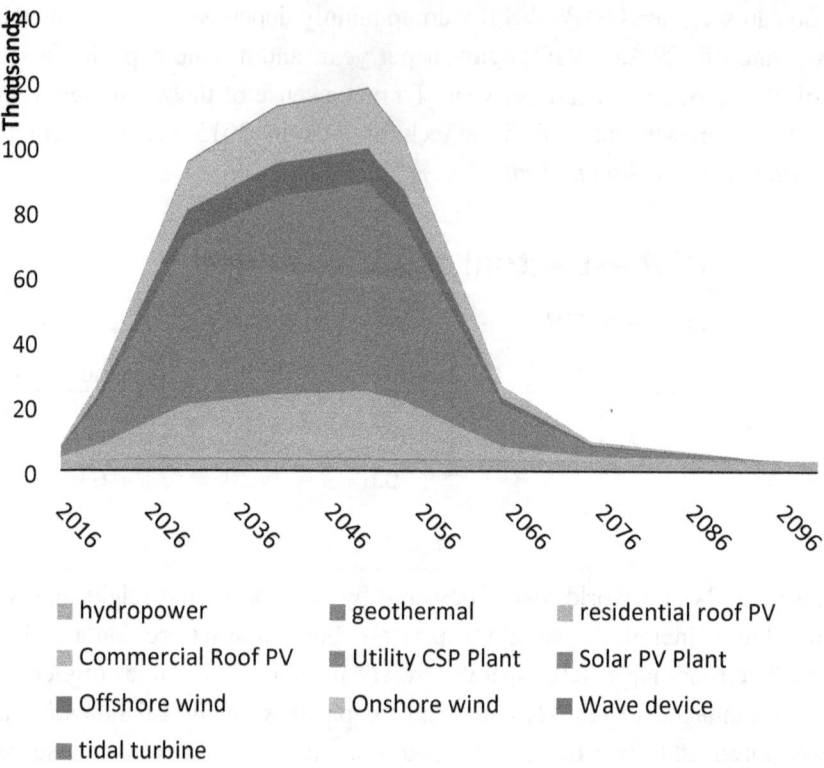

- hydropower
- geothermal
- residential roof PV
- Commercial Roof PV
- Utility CSP Plant
- Solar PV Plant
- Offshore wind
- Onshore wind
- Wave device
- tidal turbine

Figures in TWh

As you can see, the decline after the first units will be decommissioned in 2050 equals the growth curve, which means that we don't get a nice parabola like the one we would get if I mirrored the figures. However, mirroring the figures is incorrect because the decline should be as steep as the incline, from 2050 onward.

When we project the growth curve to 2100 while assuming a maintained requirement of 118,000 TWh per year until 2100, and accounting for cumulative upkeep, starting at 2050 at a rate of 100% replacements after operational lifetime, it would start to look like the graph on the next page.

The shaded part, which is called "replacements," is the cumulative upkeep that kicks in as soon as the lifetime of devices runs out, which, at that point, doubles the pressure on the production capacities to keep up with demand.

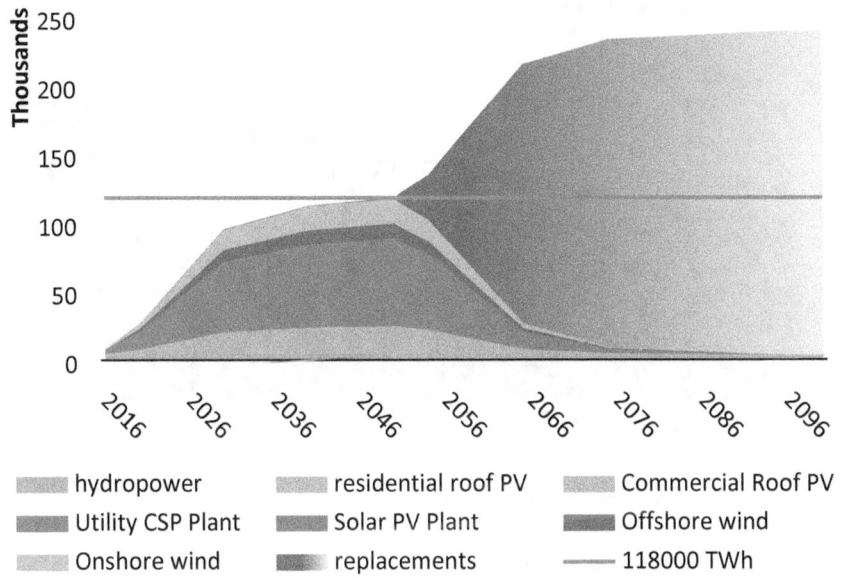

Figures in TWh

How likely is it that a growing civilization can mitigate energy consumption enough to get by on 118,000 TWh of annual WWS generation? Since we probably cannot shake off 31.4% thanks to efficiency gains, I would say that it is unlikely.

How likely is it that we can achieve this 118,000 TWh of annual WWS generation almost exclusively on wind and solar? The steepness of the curve on wind and solar additions make it seem doubtful that it is even feasible. Cumulative upkeep and replacements have been omitted, and the operational life spans of wind and solar generation have been extended in the 100%WWS Roadmap. Whichever is the case, wind and solar additions will probably be stymied by shortages of essential raw materials.

If we manage to decarbonize by 2050, do we need 118,000 TWh of electrical energy or 150,000 TWh? What about 250,000 TWh? Or even more? How much non-carbon emitting energy do we have now? What are the gaps?

According to the following graph, most growth in energy consumption is expected to be in non-OECD countries that contain 5-6 billion people. (*Organisation for Economic Co-operation and Development.*)

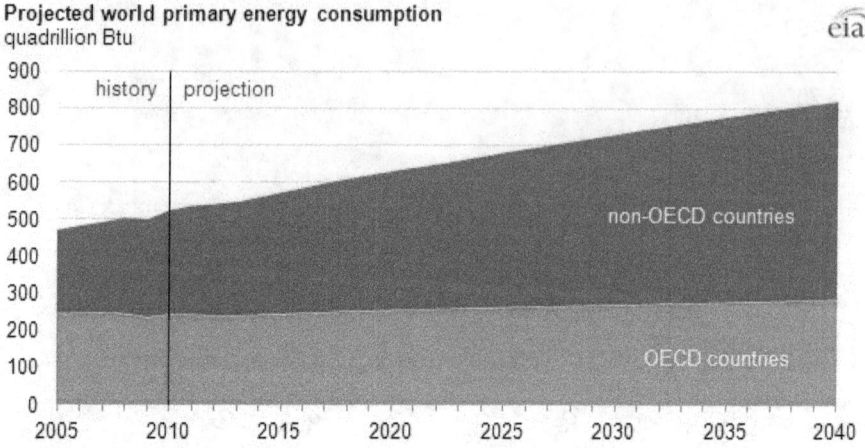

Source: http://www.eia.gov/todayinenergy/

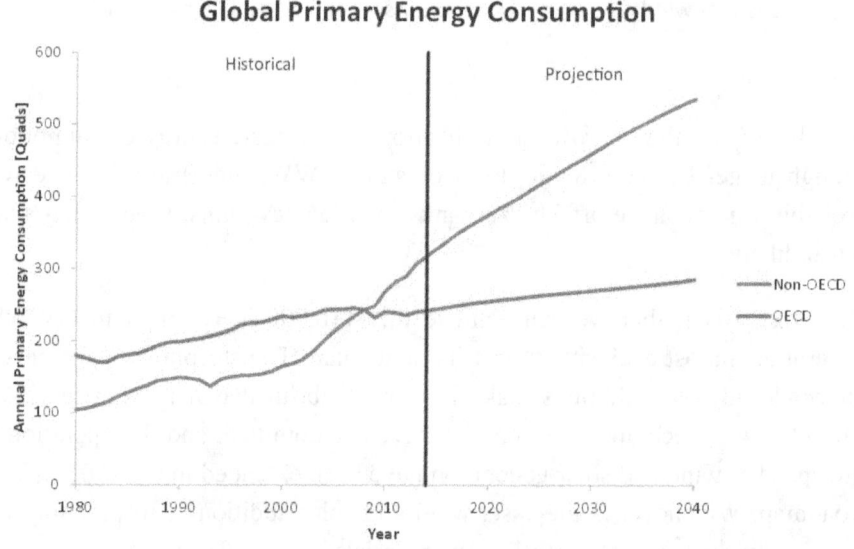

Source: http://energy.gov/sites/prod/files/2015/09/f26/Quadrennial-Technology-Review-2015_0.pdf

As you can see, the EIA and the DOE are in relative consensus where projections are concerned. Note that the DOE graph isn't stacked. If you stack the figures, they would reach the 800+ Quads mark just like the EIA graph does. (For some reason, they flipped the colors.)

If we create another graph in which we contrast what we have with what we need, our plight becomes evident—the purple bar being the equivalent of 840 Quadrillion Btu, the figure on which the previous graph ended

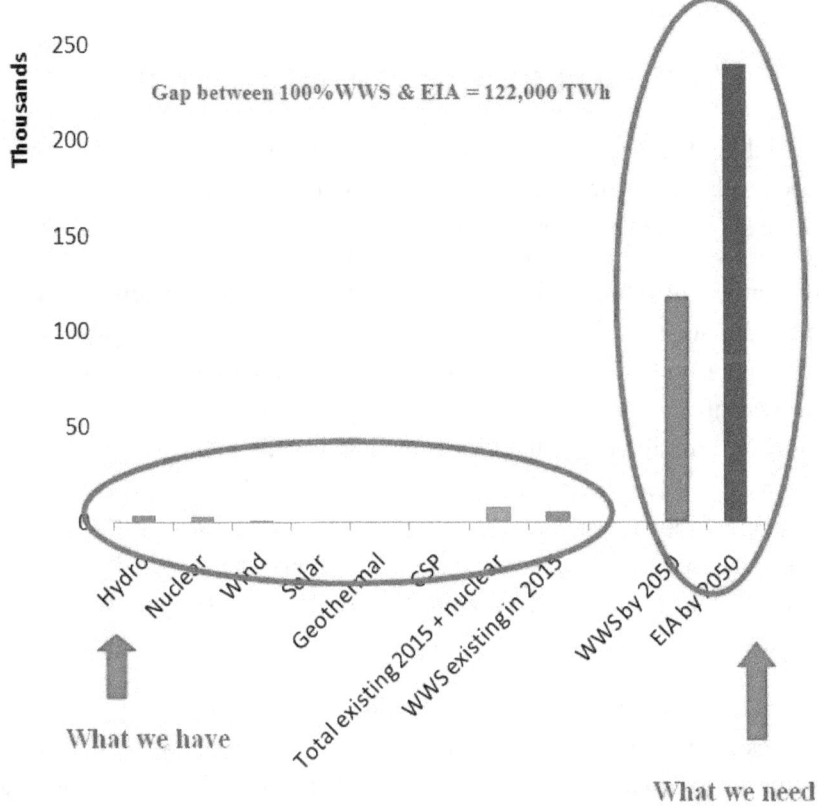

Figures in TWh

Part three: the counterargument

Debunking the 100%WWS roadmap alone is not enough, a counterargument needs to be made, for there is a powerful counterargument to be made. One which should be considered seriously.

The imminent dangers posed by climate change, as enunciated by climate scientists such as James Hansen, Ken Caldeira, Tom Wigley and Kerry Emmanuel, put immediate pressure on vital ecosystems and the long-term survival of mankind. We must decarbonize as quickly and as effectively as possible. Production capabilities of wind and solar power have to be increased exponentially in order to keep up with demand as is, let alone replacing fossil fuels. Also, it has been proven that storage for intermittent renewable sources cannot scale up quickly enough to handle the required implementation of massive amounts of renewables such as wind and solar power.

However, renewables can—and should—play an important role as power sources for processes that are intermittent as well, such as desalination, fuel synthesis, propulsion, cooling, freezing, air conditioning, water heating, and so forth. Also, renewables can play a part in helping emerging communities and economies to transition into an age of carbon-neutral energy generation. However, as demand grows and reaches levels at which baseload energy is required, a different power source is required. Enter the *verboten* "n-word": nuclear energy.

Energy reality

Reality is imposed upon us by nature in the world around us. This integral part of our existence is expressed by physics, chemistry, and mathematics.

The human population is expected to expand to roughly 9 to 11 billion people. With this expansion comes a commensurate growth in demand for water, food, and energy. The increase in energy demand and the growth of economies is already pushing the emissions of carbon dioxide and hazardous chemicals and elements (far) beyond reasonable levels.

Because renewables are marginal relative to global energy consumption, they are not helping with curbing our emissions. In fact, intermittent renewables are being offset by the addition of gas-fired power plants because no grid-scale mass storage options are presently available or economic. In addition, an increase in the production of gas-fired power will be accompanied by an increase in methane leaks, which are already high enough to negate any benefits achieved by switching from coal to carbon-reliant wind and solar alternatives.

The Aliso Canyon gas leak[65] has been deemed a natural disaster, and it is the largest methane leak ever recorded in human history, and methane as a greenhouse gas is 25 times more effective at trapping heat as CO_2.

For some reason, no gas-fired power as bridging capacity is proposed. Instead, it calls for a development of super grids and smart grids. Nevertheless, gas-fired power is the preferred method to offset renewable intermittency. Therefore, calling for increased renewable deployment without increasing grid capacity and storage, cannot deliver the benefits advertised by the 100%WWS roadmap, but it will cause a large increase in gas-fired energy production.

However, if one deems gas-fired electricity undesirable, consider biomass, which is, in effect, a euphemism for burning forests. Tree-eaters, such as Drax in the UK, consume billions of tons of shredded trees every year. biomass is one of the quickest growing "renewables" on the green end of the spectrum. In fact, if people quote high renewable growth rates, most of the time biomass is the main driver of this growth. Consider for instance Germany's energy mix. Few appreciate that biomass is the dominant source of [German renewable] energy leaving aside hydroelectricity. It's not wind, nor solar, it's burned trees... It's even more inelegant than burning coal. And worst of all, we can easily convert coal-fired power plants to run on shredded trees, as the Drax power plant in the UK has shown us. Drax used to be one of the biggest coal-fired power plants in Europe, now it is the biggest tree-fired power plant in the world. It knows no equal in terms of natural destruction. If you think that it gets fed "bio-waste" or trees that *had* to be cut down, you're wrong. Pristine forested lands have been clear cut to get the "biomass" required to fire up the furnaces for energy. *Cutting trees a la mode...* This has to stop, for it destroys nature, rather than protects it, and it is far from sustainable or green.

The possibility of runaway behavior in natural cycles is not acknowledged, and therefore vital problems remain unaddressed by the 100%WWS roadmap. These processes force us to consider geoengineering, and geoengineering doesn't come free of cost. In fact, it will require vast amounts of energy and materials, so much more that the 100%WWS roadmap will fail to provide the energy required to stabilize or slow down the negative influence of man on the planet's climate, and with it humanity.

The possibility of runaway behavior in natural cycles due to climate change is not acknowledged in the roadmap. Therefore vital problems remain unaddressed by his 100%WWS roadmap. These processes force us to consider the possible need for geoengineering, and geoengineering is very energy intensive. In fact, it will require far more energy than the 100%WWS roadmap can possibly provide.

If we examine the figures of the EIA projection of the *"World total primary energy consumption by region, Reference case, 2011-40"* we will see that primary energy consumption is expected to rise to approximately 815 Quadrillion Btu. If we take these figures, and apply the 0,6% growth rate to OECD consumption and 1,9% to non-OECD consumption and then extrapolate both figures up until the year 2100 we see primary energy consumption rise to roughly 2000 Quadrillion Btu.

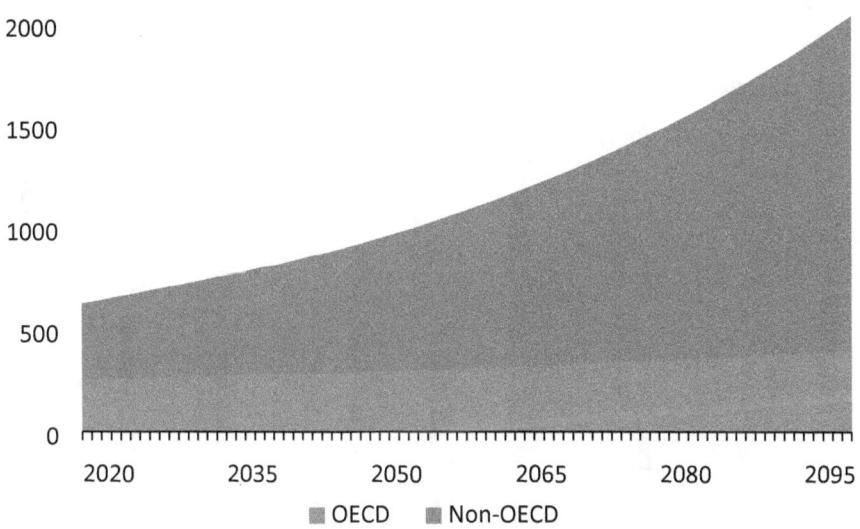

Figures in Quadrillion Btu

If we could curb the growth rate around 2050, the total energy consumption would halt at around 973 Quadrillion Btu or roughly 282,000 TWh.

Whichever of these becomes reality, is unsure. In fact, I doubt that the future will look anything like this. We simply don't know how things are going to pan out. Consider the fact that almost all of the projected growth will come from non-OECD countries. To suppose that countries—which will need great volumes of desalinated water for personal use and for agriculture and many other processes—can mitigate their energy demand is incongruent with reality. We should brace for a dramatic increase in energy consumption, not a decrease.

Even if we curtail consumption, it won't be enough. The question is "what consumption?" Is it water? Food? Goods? Fuels? And how are we to do it? We need to start working towards energy efficiency, but this isn't as easy as waving a banner that proclaims *"Renewables and efficiency will save the earth."*

Also, the growth projected in Non-OECD countries vastly outstrips anything we've seen before. It is all encompassing, growth in the consumption of water, food, electricity, energy, goods, etc. which comes in the wake of population growth and increased prosperity.

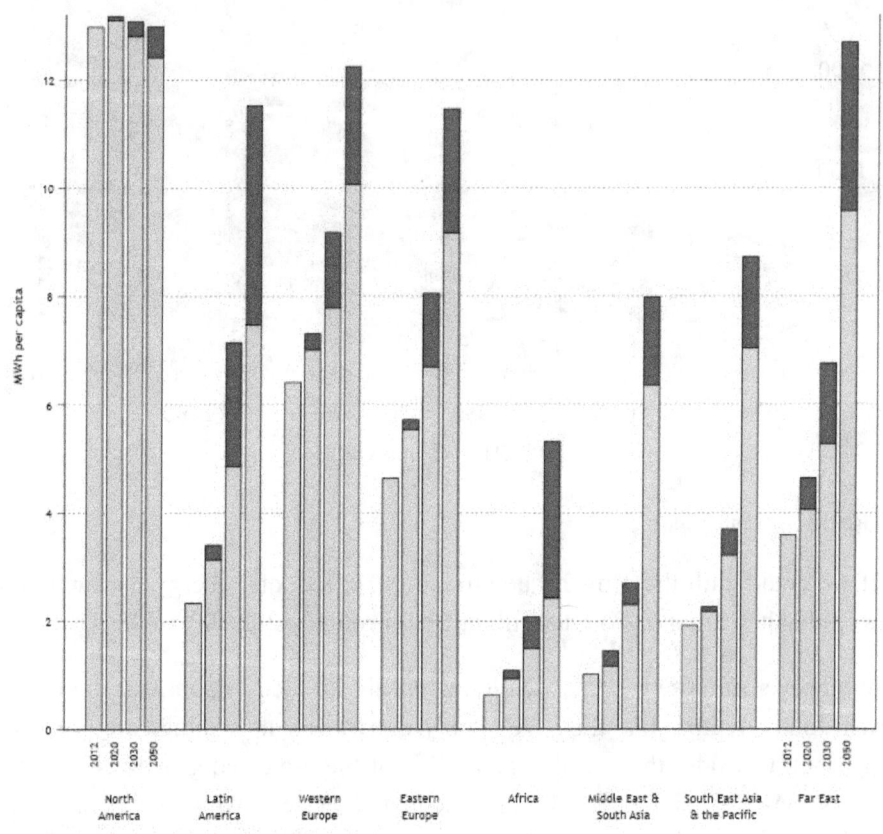

FIGURE 9. TOTAL ELECTRICITY REQUIREMENT PER CAPITA

These projections from the IAEA (International Atomic Energy Agency).[68] show that practically all regions, except for North America are expected to require more electricity per capita. Each region is segmented in four bars, the first bar represents 2012, and the subsequent bars represent 2020, 2030, and 2050. Analyses like these contradict the claim that we can cut as much energy as is proposed in the 100%WWS Roadmap.

The chart even projects growth for Western and Eastern Europe, with even greater growth projected for the Far East. And it is also important to consider the fact that Africa is trailing behind. If this model is accurate, it appears that

Latin America, the Middle East, South Asia, South-East Asia and the Pacific are going to rise to current European standards of living at a rapid pace, but will trail behind slightly when future developments are concerned. This means that they will be investing in any means of energy generation. Otherwise, this model would be impossible.

We are now going to convert the previous graphs to show figures in TWh instead of Quadrillion Btu, and I have added the growth curve required to reach the generation capacity required to successfully implement the 100%WWS Roadmap.

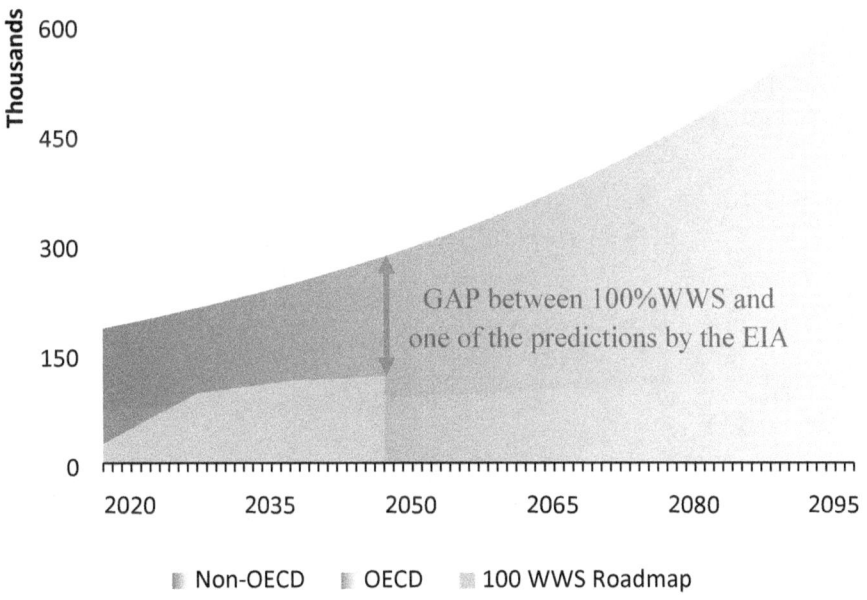

Figures in TWh

As you can see, the 100%WWS Roadmap manages to satisfy OECD consumption (at a cost of creating huge amounts of CO_2), but barely manages to get the required energy to Non-OECD countries.

The figure also doesn't reflect the requirements of upkeep. Assuming a ban on building carbon emitting power plants—which run on coal, gas or diesel—energy production would not be able to satisfy the demand, and we wouldn't be able to do any additional, essential, activities to assist the climate. We simply must account for growth.

As we reach the end of this first chapter of the counterargument, we must conclude that the 100%WWS Roadmap, which is being championed by Mark Z. Jacobson and the Solutions Project, is not an accurate model, and its predictive capabilities are inadequate. What will energy demand/consumption be in the future? Can we make it stop increasing after 2050 and keep it at a level of 118,000 TWh or 200,000 TWh or 250,000 TWh? What kind of technologies will we invent in the future? We don't know, but one thing we do know is that a very important technology has been ignored: nuclear energy.

Misconceptions about nuclear energy

Jacobson likes to embellish his 100%WWS plan by adding elements to his case seemingly aimed at discrediting nuclear energy. In fact, attempts are made to convince politicians and the public that we need absolutely no nuclear related technology at all, not even uranium mining. What is to become of medical research if we follow his plan? Must we also to shut down the flux reactors and cyclotrons required to create the isotopes for nuclear medicine and research? For instance, Bruce Power—a Canadian power company—creates the majority of Cobalt60 with CANDU Reactors. Cobalt60 is an isotope with a wide variety of important uses, including the disinfection of medical equipment. As Jacobson redounds in the idea that nuclear energy has been stagnant, he forgets that he is complicit in its decline because he keeps grasping every opportunity to pit popular sentiment against it. He thus advances as a hypothesis a partly self fulfilling prophesy. As long as he keeps helping to stymie nuclear growth by exerting his academic authority to denounce it, he remains complicit.

What happens in a nuclear chain reaction?

We can use the power that binds atoms together to create energy. Atoms stick together thanks to the strong nuclear force, brought about by the neutrons and the protons present in the nucleus of the atom. We can destabilize these forces by introducing another neutron, which may help an atom to fission or transmute. The objective of nuclear fission is for an atom to absorb a neutron and then split, making it release its energy. Our contemporary reactors do this based on solid fuel pellets which contain uranium235 (U235), a rare fissionable isotope of uranium. Within the reactor we want the process to become critical, which means that there's enough neutrons "flying" around to keep U235 isotopes fissioning and thus releasing energy and more neutrons—necessary for maintaining the chain reaction.

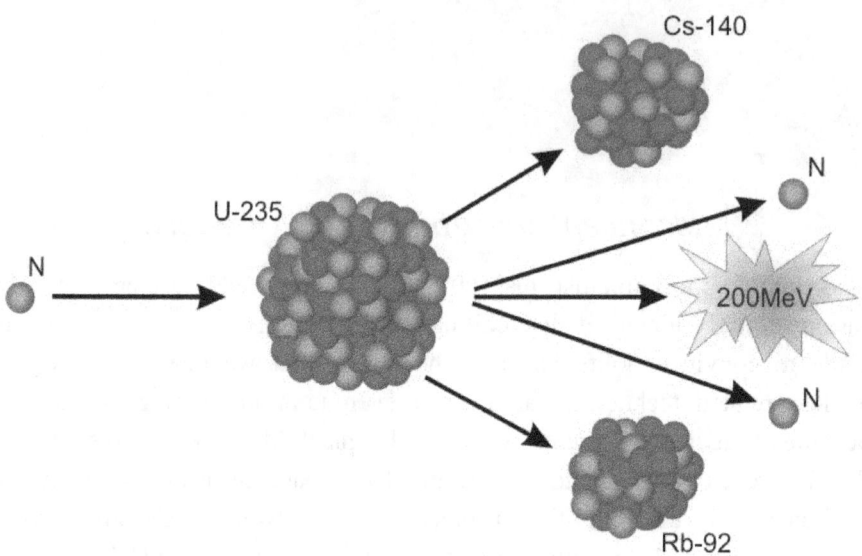

Image source: https://upload.wikimedia.org/wikipedia/commons/5/5c/Kernspaltung.png

Above you see a basic representation of what happens when one U235 atom fissions (splits). The neutron on the left is absorbed by the U235 atom, which subsequently splits into two fission products (here cesium140 and rubidium92), two neutrons, and 200 MeV (Mega electron Volt) of energy.

Only 3 to 5 percent of all the uranium atoms present in a fuel rod actually fission. Which means that 95% of the fuel remains unused. However, it cannot be used anymore because fission products have accumulated, within the fuel pellets, which absorb neutrons and therefore inhibit new fission reactions, we call these products "nuclear poisons". Basically, they eat neutrons without releasing any new ones. Also, the geometry of the fuel pellet is changed. The pellet has now been cracked, and has expanded thanks to the accumulation of fission products. The fuel for the solid fueled reactor has now been *spent*. But we need to reconsider the use of the term waste.

It's ironic that some academics focus on the problem of long-lived radioactive materials, but ignore mentioning that we have technologies that can safely consume this "waste," most of which *actually isn't waste*. Instead, most of it is fuel that has become unusable in *contemporary* reactors. And it is this spent fuel that poses the (political) issue of long-term storage.

The chemical makeup of spent fuel is such that if we reprocess it—as the French have done for decades—we can use it again. Even better, newer reactors can "burn" 90% of our stored *nuclear waste*. (It is said that there's enough energy left in spent fuel to power civilization for seventy years). Nevertheless, people keep referring to spent fuel as a problem instead of admitting that it is a potent fuel source for newer reactors. Fortunately, there are already several attempts underway to commercialize such reactors, most notably by startups Terrapower and Transatomic Power. Let's Forget about long-term subterranean storage of nuclear waste and instead make funding R&D for the reactors that are able to use spent fuel to create energy, while solving the so-called waste problem, a priority. Nuclear waste may have seemed like an afterthought, but it is the key focus for many corporations and a tremendous economic opportunity, rather than a political or environmental liability.

Another misconception perpetuated is that nuclear requires massive amounts of continuous mining operations. However, uranium and thorium mining are benign when compared to the extraction of copper, silicon and rare earth minerals. Additionally, thorium is a waste product of most mining operations. Mining corporations want to get rid of it! And we have thousands of tons of it sitting unearthed in repositories.

2014 production	Metric tons per year
Uranium	67,944
Lithium	68,768
Rare Earths	106,908
Cobalt	130,222
Silicon	7,680,000
Copper	18,435,342
Aluminium	43,478,696
Gypsum	162,589,212
Bauxite	261,896,886
Iron	1,554,479,681

Source: http://www.wmc.org.pl/sites/default/files/WMD2016.pdf
Source: http://minerals.usgs.gov/minerals/pubs/commodity/silicon/mcs-2015-simet.pdf

Somewhere between 10 and 20% of the annual silicon production is used for PV.[69] If we were to increase PV production from current levels (about 50Gw/yr) to what we will need to make the 100%WWS Roadmap happen (~1200 GW/yr), we would have to increase total Silicon production from 7,680,000 Tons per year to 26,112,000 Tons for 10% and 44,544,000 tons for 20%. A three to six-fold increase would be required within a <u>few years</u> and sustained for decades to come. Also note that the usage of Silicon for PV itself requires a twenty-fold increase, which will require continuous mining operations. Even though the roadmap assumes the implementation of circular economic principles to PV and wind technologies, it fails to reveal the volume of materials and energy required just to get started, before recycling can even begin.

If we contrast 67,944 tons of U production with a possible 26,112,000 tons of Si production, we see a sharp discrepancy between the claim that nuclear energy requires large scale continuous mining and the actual requirements for the 100%WWS Roadmap. Fuel efficiency is going to be a significant factor in this regard. Nuclear fuel efficiency is lower than 5%, which means that we leave large amounts of fuel unused, which we call spent fuel. As you have read previously, this spent fuel still contains masses of energy.

These pellets—with their energy reserves mostly untapped—will end up in spent fuel ponds and dry casket storage containers. All of this works fine; we store all of this spent fuel effortlessly.

Fortunately. we can improve nuclear fuel efficiency fifty to seventy-five fold by switching from solid fuels to liquid fuels which can yield burn-up efficiencies of 90+%. There are proven designs which have been shelved but are currently being revived by startup companies in conjunction with research institutes from all over the world.

We could double, triple, or quadruple annual nuclear additions, if we use liquid fueled power plants that can consume our stores of spent fuel and plutonium, and we wouldn't need to expand uranium mining for a long time. And even if we increase uranium production, the increase would be marginal compared to the increases for silicon and copper production required by the expansion of WWS energy as envisioned in the 100%WWS Roadmap. This demonstrates the paradoxical behavior of those who criticize nuclear innovation but fail to see the shortcomings of renewables.

We should also note that the oceans contain vast amounts of uranium that can be extracted by several processes that are close to becoming economically competitive. Our sources of nuclear fuel are ubiquitous—terrestrial and extra terrestrial—and will sustain humanity for millennia to come.

Another issue which has become a stigma for nuclear energy is its chequered history vis a vis the proliferation of nuclear arms. However, growth of nuclear energy does not necessarily imply growth of nuclear weapons. The nuclear arsenal which is in existence today has the potential to wipe out a lot of human beings with a single stroke, and it is highly immoral to keep these weapons around. What do we want to do with nuclear weapons, other than using them to scare nation-states? I would opt to dismantle them. Then what? Dismantling nuclear weapons doesn't mean that the threat will be gone because the plutonium remains, presenting a security threat. Why not use it to create energy instead? Weapons grade plutonium which was left after dismantling the nuclear arms that were decommissioned as a result of SALT and SORT has been used to generate electricity for many years in a program called *Megatons to Megawatts*.

Used properly, civilian nuclear energy is an excellent tool for de-proliferation.

As we evolve liquid fuel cycles and breeder reactors we open up pathways to remove more of these bomb-grade nuclear materials.

The risk of nuclear proliferation can further be limited by strict regulations and enforcement as countries like Indonesia, Egypt, Saudi Arabia, Tunisia, Turkey, Vietnam, Poland, and the United Arab Emirates plan to build, or are building, civilian nuclear reactors as we speak[70].

Current nuclear reactors leave minute traces of isotopes that could be used to create a nuclear bomb. These traces are locked up in spent fuel pellets, and they are hard to remove and with the correct oversight, no separation is possible. Enrichment facilities and specialized reactors which have been designed specifically to produce weapon grade plutonium are required if you want to proliferate. A case in point is that we have a wide framework of countries and organizations that watch nuclear activities closely and are ready and able to impose heavy sanctions on any country that tries to proliferate (see Iran).

Geopolitical considerations are also a factor. Consider for instance the animosity between India and Pakistan which has led India to acquire nuclear weapons. Were these a byproduct of India's civilian nuclear power program? On the contrary, India's nuclear weapons complex was a separate initiative of the Indian government involving its own military and scientific capabilities. We can gauge which countries aspire to have nuclear weapons. We knew from the outset that Iran and North Korea aspired to have nuclear weapons. The infrastructure needed to create nuclear weapons is different from the infrastructure needed to create nuclear energy for civilian purposes. The key to non-proliferation regimes is oversight—keeping a good inventory of technologies present, stockpiles, enrichment capabilities, etc.

The international community won't stop countries from developing nuclear energy for peaceful ends nor should it. Anxiety over nuclear weapons is warranted and should never go away, for it will keep us alert to their dangers, and this vigilance will serve us well in a pursuit for excellence in safety regulations and enforcement. The elimination of the nuclear energy sector as a means to stop proliferation is suggested, but this is far too simplistic and it would prevent one of our best ways to get energy from the fissionable materials in our stored waste while simultaneously greatly reducing its volume.

What about the other scourge of nuclear energy—radiation? Let us now investigate whether radiation rules out nuclear energy as a candidate for decarbonization.

First question: Is radiation dangerous?

The linear no-safe threshold (LNT) hypothesis has been used to assess radiation danger for decades. It states that any dose of radiation is dangerous and that additional doses are cumulative, meaning that the chance of developing some sort of malignancy in the body increases with each exposure. However, the LNT-hypothesis is probably wrong because it isn't supported by epidemiological evidence. It is based on experiments with an incomplete and erroneous methodology and hasn't been retested for years. Corroboration is the key to establishing its validity, but up until now knowledge in this regard is limited.

The counter-hypothesis, however, is supported by epidemiological evidence, and it is called Hormesis[71]. Hormesis describes a model of physiological responses to radiation doses that are beneficial within certain bounds. It is hypothesized that small amounts of radiation activate repair mechanisms in biological cells.

Second question: Where does radiation come from?

From high school physics the reader may recall that atoms have a nucleus of neutrons (except for hydrogen1) and protons and a "cloud" of whirring electrons. The protons are positively charged and repel each other, and this means that the nucleus of an atom cannot exist if it has two or more protons and no neutrons, it would simply fly apart. The neutrons are there to counter the repelling forces, and the force it brings to bear is called the strong nuclear force. Neutrons make the nucleus of atoms stick together and keep it stable as long as possible. And that's the catch, as long as possible. If the nucleus of an atom is unstable, it loses mass and energy over time, and this loss causes the emission of different particles. The most important ones are called Alpha, Beta, and Gamma. The amount of neutrons and protons and electrons can/will change during radioactive decay.

The clue here is the decay time that is associated with each different isotope. A short decay time means that the atom is highly radioactive; A long decay time like that of uranium means that the atom is only slightly radioactive. Therefore—and this might seem like a contradiction—it is not dangerous to hold a piece of pure uranium in your hand.

Radiation is ubiquitous and omnipresent. The Earth is constantly being bombarded with cosmic radiation that comes from the nuclear fusion reactions in the sun; We breathe radioactive radon and radium; We consume minute traces of radioactive potassium because it is in our food; Our granite counter tops are radioactive due to the presence of thorium and uranium. Even our bodies are naturally radioactive.

We can use the Aspirin analogy here: as everyone knows, we can kill ourselves by ingesting a ridiculous amount of aspirin; however, if you have a headache, an aspirin or two can provide harmless pain relief.

This also applies to radiation. Too much at once can certainly kill you. But sustained exposure to low levels of radiation, which we constantly experience, may be beneficial, rather than deadly, as demonstrated by hormesis, which disputes LNT. It is also important to note that nuclear power plants are highly regulated and are required to contain every speck of radiation. And if there's an accidental release of radioactive elements, it hardly exceeds natural radiation levels and quickly becomes very dilute.

Fear of radiation has long been used by those who know that nuclear power can put an end to carbon-fueled power plants, and many environmentalists use it to promote wind and solar alternatives—which greatly rely on fossil fuels to offset their intermittency. Unreasoned fear is the largest hurdle that civilian nuclear energy must overcome.

Third question: What about Chernobyl and Fukushima?

It is true that these accidents. Which were far more frightening than damaging, have had a profound influence on the public perception of nuclear power. In fact, UNSCEAR and the World Health Organization cannot corroborate the high numbers of cancer incidence predicted by proponents of LNT. So what are the numbers?

Chernobyl : confirmed 50 dead, 4000 non-fatal cancers (UNSCEAR)

Contrast those with the following annual worldwide death-tolls.

Smoking : 6 Million of which 600 000 non-smokers (WHO)
Alcohol : 3.3 Million in 2012 (WHO)
Car-accidents : 1.25 Million in 2013 (WHO)
War : 180,000 in 2014 (IISS)
Malnutrition : 3.1 Million children per year (WFP)
Malaria : 438,000 deaths (WHO)
diarrhea : 1.5 Million (WHO)

As you can see, there are more important causes of death to start working on. In fact, trying to mitigate deaths from Diarrhea and malnutrition will cost us immense amounts of energy.

As of November, 2016, the figures for Fukushima stand at zero. Even if these reliable agencies somehow missed perhaps 10,000 deaths from Chernobyl or Fukushima, that number is insignificant to the deaths caused by the combustion economy or by drinking, smoking and driving. However, the claim of 1,000 deaths from Fukushima is not supported by epidemiological evidence. Perhaps the fact that these deaths were caused by the evacuations will be pointed out, but then it begs the question whether it was the power plant or the fear of it that killed these people?

Fear of radiation comes from sensationalism—not from science and empiricism. Ask any radiologist, oncologist or any doctor if they are afraid of ionizing radiation? If LNT was true, cancer research—and the researchers, one of whom is a good friend of mine—would be dead in the water, and there would be litigation without end.

If LNT were true, we could sue the coal industry for emitting traces of uranium and thorium. we could sue Exxon for fracking because it releases radioactive Radon and Radium, and we could sue the airlines for not telling us that flying exposes us to higher doses of background radiation.

We now know that a nuclear accident is dangerous, but its consequences are primarily economic, political, and social. We have learned how to prevent them. In the many thousands of hours of nuclear power plant operations, only Chernobyl caused any deaths. Generation III and IV reactors are even safer due to the addition of passive safety features that require no human input.

What about Lifecycle emissions?

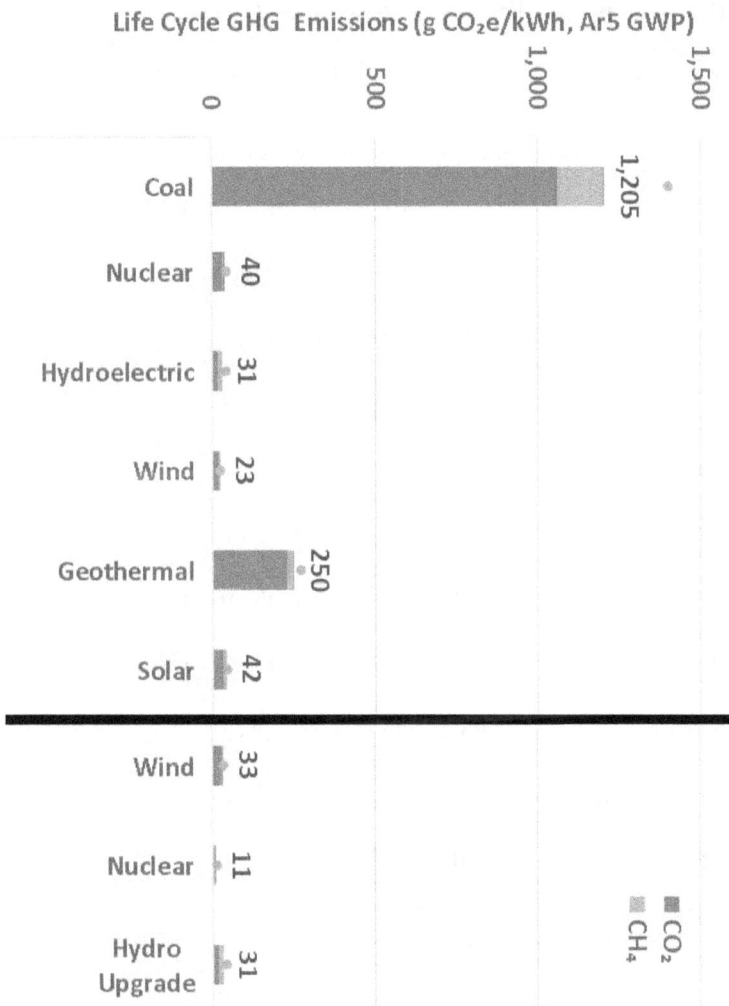

source: https://www.eia.gov/conference/2015/pdf/presentations/skone.pdf

The figures above the black bar constitute lifecycle emissions as of 2013. those below the bar indicate projected emissions after technological advances have been made. As you can see, nuclear lifecycle emissions are as low the mainstream renewables. Also, note the stark contrast between nuclear and coal. I wonder, though, why wind became worse. Perhaps it is because of added maintenance and decommissioning.

What about deaths per unit of energy generated?

Energy Source	Mortality Rate (deaths/trillion KWh)
Coal – *global average*	100,000 (50% global electricity)
Coal – China	170,000 (75% China's electricity)
Coal – U.S.	10,000 (44% U.S. electricity)
Oil	36,000 (36% of energy, 8% of electricity)
Natural Gas	4000 (20% global electricity)
Biofuel/Biomass	24,000 (21% global energy)
Solar (rooftop)	440 (< 1% global electricity)
Wind	150 (~ 1% global electricity)
Hydro – global average	1400 (15% global electricity)
Hydro – U.S.	0.01 (7% U.S. electricity)
Nuclear – global average	90 (17% global electricity)
Nuclear – U.S.	0.01 (19% U.S. electricity)

Figures by James Conca, excerpted from Forbes.com.
Source: http://www.forbes.com/sites/jamesconca/2012/06/10/energys-deathprint-a-price-always-paid/#59f51f849d22

Included in these figures are deaths caused by Chernobyl and Fukushima. So when we put this into context, we have to conclude that nuclear energy is a safe source of energy.

Also, consider this excerpt from Pushker Kharecha's publication *"Coal and Gas are Far More Harmful than Nuclear Power"*[72]

"We conclude that nuclear energy — despite posing several challenges, as do all energy sources (ref. 7) — needs to be retained and significantly expanded

in order to avoid or minimize the devastating impacts of unabated climate change and air pollution caused by fossil fuel burning."

All we have done in this chapter is try to establish whether or not it is justified to shut down nuclear power in order for a 100%WWS Roadmap to be implemented. Since the increase in wind and solar must be accompanied by similar increases in silicon and copper production we may conclude that the 100%WWS Roadmap is almost impossible to implement. However, why is it necessary to stop nuclear power anyway? The claim that nuclear has high carbon dioxide emissions during its lifecycle has been proven to be wrong, and therefore nuclear energy should be a key focus in a future of energy precisely because it is low-carbon and because it has a big punch and therefore is least damaging to the environment.

Given the high probability that renewables like wind and solar are unable to scale as advertised it logically follows that we need additional sources of non-carbon emitting energy. Additionally, current nuclear facilities have life spans of fifty to eighty years, and once [small] modular reactors become a main fixture in the world of energy—and they will—we will see that nuclear facilities can stay in operation far longer albeit with routine maintenance and overhauls and part replacements. Unmatched longevity is an important "ace in the hand" of "cards" possessed by nuclear power.

Most importantly, we are now going to examine the potential of nuclear technologies to bring about the obsolescence of fossil fuels. The basic premise is this: With its high capacity factors and the high energy yield per unit of materials invested, nuclear is the best possible way to defeat coal, gas and oil.

The "crown jewels" in the nuclear argument are Sweden, France, and Switzerland. These countries have literally extirpated fossil fuel electricity generation by replacing them with nuclear energy and hydropower. The beauty of a nuclear facility is that it immediately offsets multiple coal-fired power plants. For instance, if you have six 1000 MW coal-fired power plants, you'd need four 1000 MW nuclear power plants to offset their generation potential i.e. roughly 32 TWh. Subsequently, if we look at gas-fired power plants, the discrepancy becomes even bigger. The main driver of these differences is the factor at which the capacity of said power plant is actually used.

The folly of Germany's energy transition has paradoxically been advertised as a great success for the mass implementation of renewable energy sources, but is it? It is slowly, but surely, dawning on the Germans that their choice was wrong and immoral. Why would it be immoral? They chose to close down their nuclear capabilities and knew that, as a consequence, they had to increase the consumption of brown coal, which has been proven to kill people with its noxious and hazardous emissions. And now, it has become clear that they need to keep open their coal-burning facilities until at least 2040. It is inexplicable to replace nuclear power with coal, as the Germans have done. Whilst it is true that they also added a huge number of PV panels and wind turbines, on a macro scale, they amount to little. Claims are made regularly that Germany gets more than 80% of their energy from renewables. This is a lie. First of all, such claims refer to electricity and not energy. For such claims to be true in terms of total energy most Germans would have to keep their cars and trucks in the garage. If they ever get 80% of their electricity from renewables, it is only remotely possible on a windy, yet sunny, holiday.

Is nuclear energy too expensive?

It depends on the situation, the design, the country and the regulations. Here is what experts like Jessica Lovering, Arthur Yip and Ted Nordhaus have to say:[73]

"This paper presents a new data set of historic experience curves for overnight nuclear construction costs across seven countries. From these data, we draw several conclusions that are in contrast to the past literature. While several countries show increasing costs over time – with the US as the most extreme case – other countries show more stable costs in the longer term and cost declines over specific periods in their technological history. Moreover, one country, South Korea, experiences sustained construction cost reductions throughout its nuclear power experience. The variations in trends show that the pioneering experiences of the US or even France are not necessarily the best or most relevant examples of nuclear cost history.

These results show that there is no single or intrinsic learning rate that we should expect for nuclear power technology, nor an expected cost trend. How costs evolve over time appears be dependent on different regional, historical, and institutional factors at play. The large variance we see in cost trends over time and across different countries – even with similar nuclear reactor technologies – suggests that cost drivers other than learning-by-doing have dominated the cost experience of nuclear power construction. Factors such as utility structure, reactor size, regulatory regime, and international collaboration may play a larger effect. Therefore, drawing any strong conclusions about future nuclear power costs based on one country's experience – especially the US experience in the 1970s and 1980s – would be ill-advised."

What about putting people to work?

In terms of creating jobs, nothing beats wind and solar, but is this a good thing? Ripudaman Malhotra wrote an excellent blog post on the efficiency of energy technologies as a function of per capita workforce. I discovered that carbon-free nuclear power creates the fewest jobs/kWh, and carbon-reliant wind and solar, the most. we see here an inverse relationship. In the roadmap it is assumed that we would create roughly 50 million jobs over 35 years in construction and operation, while the lost nuclear workforce (including mining) would total some 1.15 million jobs, effectively shutting nuclear down.

The following material comes from the REN21 2015 Global Status Report (Table 1), the REN21 2016 Global Status Report, Key Findings (Page 9), the 100%WWS Roadmap per April 24. 2016 (Table 10), the EIA Electric Monthly, July 2016, and World Nuclear Power Reactors & Uranium Requirements—by the World Nuclear Association. All figures are 2015 & global.

Sector	# Workers	Energy Produced (TWh)	Productivity (GWh/in industry employed person)
Nuclear	1,150,000	2,441	2.12
Wind	1,027,000	1,233	0.99
Solar	3,281,000	569	0.17

Suppose we implement the 100%WWS Roadmap with 29,156,000 GWh for wind and 81,851,000 GWh for solar (together roughly 111,000 TWh), we'd be adding 29 million jobs for wind and 471 million jobs for solar, but Given our current economic models (and total population) this is completely unrealistic.

To generate 250,000 TWh per year, we will need 117 million people in the nuclear industry, which far less than the 1.4 billion required for solar or 252 million for Wind, but it makes hands free for other, equally important, jobs like healthcare, farming, engineering, research, and community service. It has also been proven that the deployment of nuclear energy is beneficial to the GDP of a country. It is efficient, and, therefore, the best choice and the greatest enabler.

Claims are made that we would "just" 50 million jobs for wind and solar, but current productivity figures don't support this assertion. Whichever path we take will shift many jobs from the fossil fuel industry to nuclear, wind, solar, and the other non-carbon energy technologies.

When contrasted correctly, nuclear energy outperforms any other technology in terms of energy generation per unit of material invested, per unit of money invested and per the numbers of workers required. Only nuclear power has demonstrated operational excellence, unsurpassed safety records, and minimal damage to the environment.

Almost all arguments against nuclear energy are specious, and the efforts of anti-nuclear activists are damaging our ability to effectively counter anthropogenic climate change.

Groundbreaking nuclear innovations

The nuclear energy industry has begun to revolutionize itself with a plethora of superior new designs. This does not detract from the immense value our current reactors have delivered over their operational life spans, and still do, today. The expansion of nuclear capabilities will have a multitude of positive spin-offs. Consider, for instance, the demand for engineers, nuclear scientists and maintenance workers, the increased availability of reactor-produced medical isotopes and the isotopes needed for space exploration. Nuclear power has been a dominant force for good in this world.

How are we going to revolutionize the world of nuclear power? First, we must standardize all new designs for light water reactors, breeder reactors, molten salt reactors and various other modular reactors. Moving from today's generation II reactors to generation III+ and IV reactors will be a quantum leap in terms of deployment speed, fuel efficiency, costs, and safety. In order to make a rapid deployment possible, reactor facility production and building speeds must be augmented significantly. Two important developments will make this possible. These aren't far-fetched ideas, but *simple* adaptations of principles that have been with us for decades, if not almost a century. The assembly line heralded by Henry Ford in 1913 is one of these principles and from this, we automatically arrive at the design principle of modularity.

Additionally, consider the design of contemporary coal-fired power plants. Their generators don't need to rely on thermal energy from burning coal. We could easily demolish the coal-burning section and drive the generators with steam from nuclear power. In other words, we could build the reactor facilities and use the existing generators to replace several thousand of coal-fired power plants all over the world. It would save construction time and money. But it would also help us clean up the areas that have been contaminated by decades of coal production. We could use excess heat from reactors to clean up tailings like coal ash heaps, thus diminishing the environmental impacts of burning coal. This proposal is rarely considered by those who fantasize about 100% renewable futures.

Cogeneration—using excess heat from the reactor which would otherwise be expelled—can easily provide district heating, seawater desalination, pulp & paper manufacture, synthesizing fuels, or even improving the efficiency of blast furnace steel making and other thermal processes of the suchlike. neither wind nor solar can be used for cogeneration. Desalinating seawater will become very important for alleviating stress on fresh water sources, and nuclear power can supply the required power. Normally, waste heat from energy generation is expelled, but we could use this waste heat for desalination and other processes.

Nuclear power has the highest efficiency in terms of materials invested per unit of energy generated; it has the highest efficiency in terms of fuel invested per unit of energy generated; it has the possibility to provide excess heat for other processes; it has the highest efficiency in terms of longevity, and it provides long-term stability in terms of economics, jobs, and energy security. However, we have to point this out *ad infinitum*, because of the continued opposition to nuclear energy.

The 450 civilian nuclear reactors that are currently operating chiefly consist of Generation II reactors, some of which will have to be replaced within a few decades, primarily those that are sixty years old. Many people tend to scorn these older reactors, but they ignore the fact that these reactors have created no carbon dioxide and kept coal-fired power plants from being built. In fact, Sweden, Switzerland, France and Belgium have replaced carbon emitting power plants in favor of nuclear energy, which has helped them almost completely decarbonize their electricity grids.

However, the people who influence the public against nuclear power claim that the industry has been stagnant, but that is no longer true. More than 50 small modular reactor (SMR) designs[74] are being developed. SMRs usually deliver about 300 MW of electricity, can be mass produced, and can be deployed much faster than contemporary designs. In addition, two generation III+ reactor designs are currently being built, namely the AP1000 and the EPR (European Pressurized Reactor).

The AP1000 is the first generation III+ reactor to go online. Unlike current reactors, generation III+ models have *passive* safety features, modularity, and a very high degree of standardization. China is currently investing most of their effort in nuclear energy and have set a target of at least 60 new nuclear power plants[75] over the coming decade. It will take Chinese builders roughly eight years to complete the first AP1000 ever built, and this construction time is expected to eventually drop to four years. generation IV reactors combine high fuel efficiencies with high temperatures and "walk-away" passive safety features.

The AP1000 will probably dominate the industry for a couple of decades because it is currently the best pressurized water reactor and it promises high deployment. However, if we deploy thousands of these, we might increase the demand for uranium too much. The materials and feedstock issue remains, and we should be provident.

Below you see a cross section of a light water reactor.

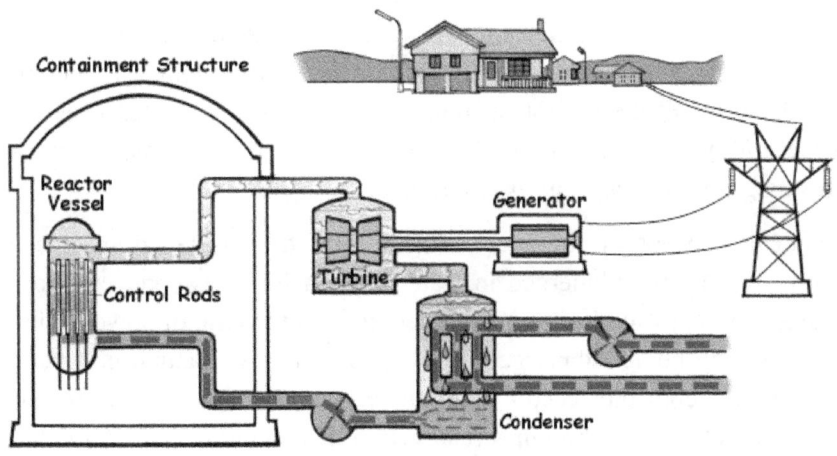

Image credit: http://www.nrc.gov/reading-rm/basic-ref/students/animated-bwr.html

And here is a Pressurized Water Reactor.

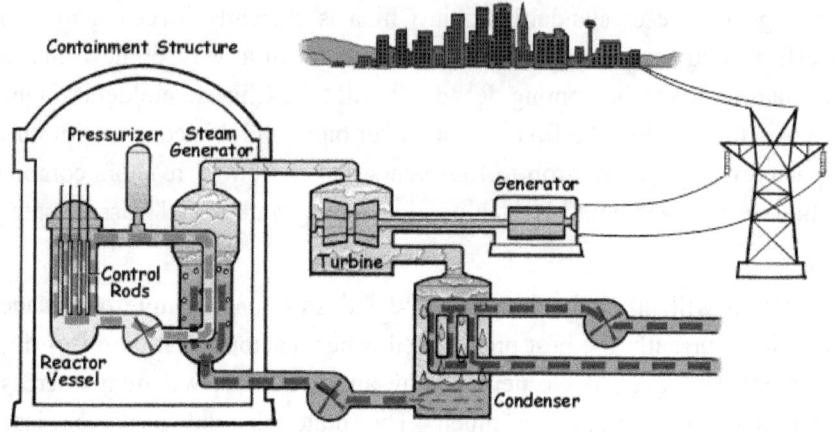

Image credit: http://www.nrc.gov/reading-rm/basic-ref/students/animated-pwr.html

These images portray the nuclear designs in use today. Even the AP1000, which is the most modern iteration, is a pressurized water reactor, which uses a solid fuel cycle. SMRs like the *NuScale* design by NuScale Power, also rely on these principles. If you scale down light water reactors to a capacity of 300 MW or less, you have a small modular reactor. The modularity pertains to manufacturing of the reactors as plug-able modules that creates scalability with the ability to "stack" multiple reactors in one facility, expanding, or decreasing generation capacity when needed.

The NuScale concept allows up to twelve reactors to be installed in a special reactor-pool. All are submerged in water, and each one is connected to its own generator. NuScale Power is confident that, in the case of power failure, heat removal by being submerged in water, and through natural convection will be able to cool the reactor indefinitely. Because the NuScale Reactor contains very little fuel, it requires relatively little cooling in case of a failure. Also, the size of the reactor permits factory manufacture and road, rail, or waterway transportation of "off-the-shelf" reactors.

Perhaps interesting to note is the progress on pebble bed technology being made at Berkeley University, these reactors feature other passive safety features. The fuel for these reactors is embedded in graphite balls. An operational PBR exists in China. An interesting iteration in the SMR field.

These designs offer great flexibility and allow nuclear power plants to operate much longer than those we use today because the modules can be exchanged when needed.

For more on SMRs, please download the *2016 Edition, Advances in Small Modular Reactor Developments, by the IAEA* (International Energy Agency).

There is also a commercial *breeder* reactor in operation, a form of reactor long believed to be elusive following the "failures" of the EBRI & EBRII, the FERMI, DFR, PFR, Phenix, Superphenix, and the Monju power plants. The BN-800 in Russia is the only liquid metal (sodium is a metal) fast breeder reactor in operation today. What is so special about [fast] breeder reactors? They make more fuel than they initially get, and they can burn actinides that would otherwise remain unused. The main fuel, in the case of the BN-800, is plutonium bred through the transmutation of ubiquitous uranium238.

The Advanced Fuel CANDU Reactor (AFCR) is a heavy water breeder reactor being developed in Canada and China.[76] It is flexible in fuel usage and can incorporate thorium as a fuel. Other interesting reactors like GE-Hitachi's PRISM reactor are designed to eliminate the existing *"plutonium problem"*.

About MSR's

Image: The first MSR ever in operation at Oak Ridge National Laboratories, in the 1960s.
(Note the man with the wrench on top of the Reactor...)

Four MSR Designs are particularly interesting, two of which are descendants of the original MSRE which ran in the late 1960's; one of which is being marketed as a nuclear waste burner; and one as the ultimate breeder reactor. The names of the companies behind these designs:

- Terrestrial Energy (the Integral Molten Salt Reactor, IMSR)
- ThorCon Power (the ThorCon reactor, MSR)
- Transatomic Power (Transatomic Power MSR, TAP MSR)
- FliBe Energy (the Liquid Fluoride Thorium Reactor, LFTR)

The true breakthrough in nuclear energy will be found within the realm of Generation IV designs, one of which is my personal favorite, the brainchild of Alvin Weinberg, the molten salt reactor (MSR). The MSR is actually quite an old design, it has been around since the nineteen-sixties. A very challenging question spurred Weinberg wanted a compact and lightweight reactor design that didn't require a separate cooling loop and that could be used to power a bomber without using solid fuels. The result was a reactor in which the fuel was dissolved into a molten salt, which was both the working fluid and the coolant simultaneously.

Nominally the selected salt has a melting point somewhere around 400 degrees Celsius, which means that if the temperature drops below 400 degrees Celsius it solidifies, and this is an excellent characteristic for a fluid that is simultaneously the working fluid and the coolant. This means that if there's a loss of power the fluid can be drained into a special drain tank, which has no moderator, and this means that the fission process in the fluid will stop and the fluid itself will cool down and solidify—until someone heats it up again. This process removes the fear for meltdowns because the solidified salt/fuel combination cannot melt until heat is deliberately supplied. This characteristic makes a sudden shutdown of the reactor a "non-event" when compared to our current reactors that require constant cooling processes to prevent core temperatures from rising too high.

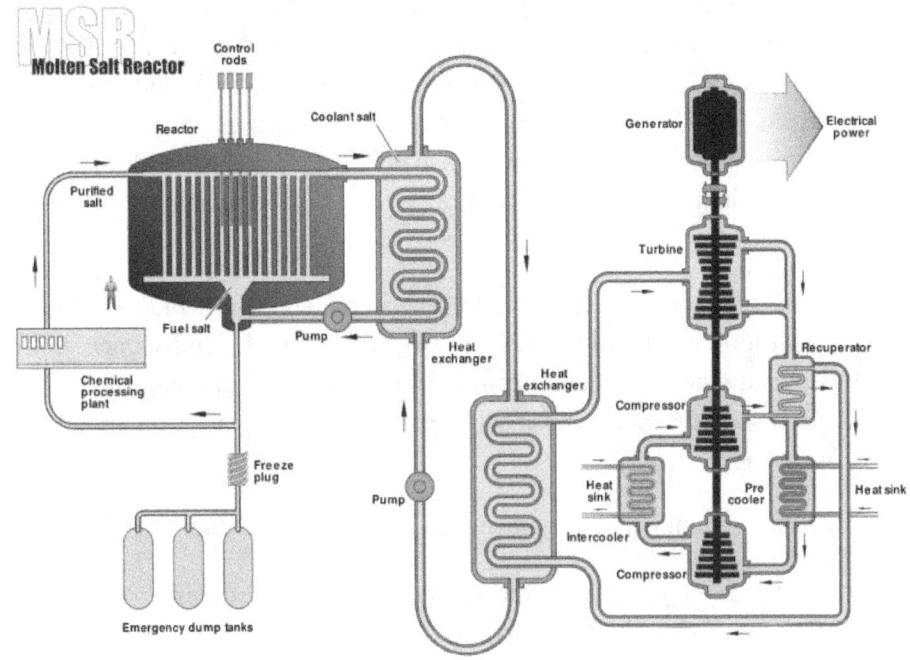

How do MSRs differ from our light water reactors?

LWR/PWR fuel consists of solid fuel pellets that, unfortunately, trap fission byproducts that prevent further chain-reactions. As a consequence, fuel efficiency is low, and what some call "waste" is high, even though this "waste" is actually untapped nuclear fuel.

MSR fuel is a liquid. The uranium (or thorium) is dissolved in a molten salt, which means that it can move around freely, and that fission products can be extracted, which, improves fuel efficiency from 3~5% in an LWR/PWR to about 95% in an MSR. Because of this improvement in efficiency, MSRs can provide energy not just for a few centuries but for thousands of years.

MSRs can utilize a variety of salts plus several uranium isotopes (not just the rare U235), and fertile elements like thorium232, which offer us the possibility to extract valuable (waste products) isotopes like bismuth213 or isotopes required for space exploration.

LWRs and PWRs get cooled by water, but this water must be pressurized to keep it from boiling, which it normally does at 100 degrees Celsius. However, high pressure means that if there's a break anywhere, the pressurized, superheated water that escapes will gasify immediately, causing its volume to increase exponentially. Therefore, a LWR/PWR requires a special containment dome to contain any events—however improbable they may be.

The MSR, however, operates at atmospheric pressure, so if there's a rupture in the fuel loop there will be no blow-out, the fuel will solidify once it exits the loop, and there's no requirement for a containment building. The MSR is the best design available, and it will revolutionize nuclear power within the coming decade.

The most important general characteristics of the MSR design are these:

- Successfully tested and proven in the 1960's for multiple years at Oak Ridge National Laboratory
- Production of valuable isotopes
- Factory / mass production of all critical components
- Fully standardized and modular designs
- Core swapping capabilities / enhanced operating lifecycles
- Scalable
- Fast and high volume deployment figures
- Ability to cogenerate energy for
    - Desalinated water
    - Community heating
    - Fertilizers
    - Concrete & bricks
    - Thermoplastics
    - Blast Furnace Steelmaking
    - Chemical Synthesis
- Significantly higher fuel efficiencies over solid fuel reactors
- Ability to utilize far more different and ubiquitous fuels, including nuclear waste and weapons-grade plutonium
- Very low waste profile compared to contemporary reactors

- Waste profile also less long-lived, down to 300 years maximum, in a manageable volume
- Fuel-salt versatility
- Possibility of using either a burner or a breeder design
- Stretches fuel cycle capabilities into the thousands, millions, perhaps even billions of years on terrestrial resources, let alone extra-terrestrial resources.
- Operates at high temperatures, but low, almost atmospheric pressures
- Passive safety through high freezing point, freeze valve, and automatic, non-engineered, heat removal capabilities
- Probably cheaper than coal
- Highest possible ERoEI (Energy Returned on Energy Invested)

Terrestrial Energy - the IMSR

The IMSR is being designed by Terrestrial Energy. Terrestrial Energy probably has the best talent pool in the MSR sector, and they are forging ahead at a pace that leaves me confident that they will reach their intended targets and eventually will commercialize their design. Will it be the 2020's? It is certainly what they are aiming for. Here are some of the unique selling points as advertised on / and excerpted from their corporate website:

Cost

- The IMSR can produce power at a Levelized Cost Of Energy of $0.04 to $0.05/kWh
- The IMSR capital costs are competitive with fossil fuels
- The IMSR operating costs are materially lower than conventional nuclear

Scalability

- The IMSR is designed in multiple power outputs, from 80 MWth (29 MWe) to 600 MWth (300 MWe)
- The IMSR can be factory-built and transported by truck or rail
- Scalability allows the IMSR to be used in industrial processes exclusively fuelled by fossil energy

Accessibility

- The IMSR has a small land footprint that will not encroach on the natural landscape
- The IMSR does not require water to operate
- The IMSR is grid-independent energy at the point of demand

The IMSR bears all the hallmarks of a game-changing technology, and this is continuously being underlined by frequent additions of backing from prominent corporations like Caterpillar, NB Power, and Duke Power. Additionally, Terrestrial has been successful in establishing good working relationships and partnerships with Oak Ridge National Laboratories, the Canadian Nuclear Safety Commission, the Dalton Nuclear Institute, the University of Tennessee, and the Canadian Nuclear Laboratories. Top industry professionals keep being affiliated with Terrestrial Energy, high ranking people who used to work for Goldman Sachs, Lockheed Martin, Westinghouse, the EPA, Bechtel Nuclear, Atomic Energy of Canada, and so on and so forth. The fact that Terrestrial Energy has been so successful in engaging these world-class industry professionals is a testament to the quality of their vision and the obvious feasibility of their design.

It is hard not to go into hyperbole, but this company has the potential to revolutionize civilian nuclear energy, and probably will, somewhere in the mid-2020's.

The relatively small IMSR (which still can reach up to 300 MW per unit) is transportable over the road, rail, and waterways. Its components are designed to be modular and thus to be produced in a factory, and this possibly ensures a high output of reactors per year, meaning that it can play a significant role in the decarbonization process. It is likely that they will be able to produce somewhere between 50 and 100 reactor units per year, perhaps even more, but first they have to pass all the legislative hurdles and build their commercial demonstration power plant, which is based on the successful MSRe at Oak Ridge National Laboratories, back in the 1960s.

The IMSR will be a closed molten salt reactor which utilizes low enriched uranium.

ThorCon Power - the ThorCon reactor
Excerpted from ThorConpower.com
Safe

ThorCon is a simple molten salt reactor. Unlike all current reactors, the fuel is in liquid form. If the reactor overheats for whatever reason, ThorCon will automatically shut itself down, drain the fuel from the primary loop, and passively handle the decay heat. There is no need for any operator intervention. In fact there is nothing the operators can do to prevent the drain and cooling. **ThorCon is walkaway safe.**

The ThorCon reactor is 30 m underground. ThorCon has three gas-tight barriers between the fuel salt and the atmosphere. Two of these barriers are more than 25 m underground. Unlike nearly all current reactors, ThorCon operates at near-ambient pressure. In the event of a primary loop rupture, there is no dispersal energy and no phase change. The spilled fuel merely flows to a drain tank where it is cooled. The most troublesome fission products, including strontium-90 and cesium-137, are chemically bound to the salt. They will end up in the drain tank as well.

No New Technology

ThorCon is all about NOW. ThorCon requires no new technology. ThorCon is a straightforward scale-up of the successful Molten Salt Reactor Experiment (MSRE). The MSRE is ThorCon's pilot plant. There is no technical reason why a full-scale 250 MWe prototype cannot be operating within four years. The intention is to subject this prototype to all the failures and problems that the designers claim the plant can handle. This is the commercial aircraft model, not the Nuclear Regulatory Commission model. As soon as the prototype passes these tests, full-scale production can begin.

Rapidly Deployable

The entire ThorCon plant including the building is manufactured in blocks on a shipyard-like assembly line. These 150 to 500 ton, fully outfitted, pre-tested blocks are barged to the site. A 1 GWe ThorCon will require less than 200 blocks. Site work is limited to excavation and erecting the blocks. This produces an order of magnitude improvements in productivity, quality control, and build time. ThorCon is much more than a power plant; it is a system for building power plants. A single large reactor yard can turn out one hundred 1 GWe ThorCons per year.

Fixable

No complex repairs are attempted on site. Everything in the nuclear island except the building itself is replaceable with little or no interruption in power output. Rather than attempt to build components that last 40 or more years in an extremely harsh environment with nil maintenance, ThorCon is designed to have all key parts regularly replaced. Every four years the entire primary loop is changed out, returned to a centralized recycling facility, decontaminated, disassembled, inspected, and refurbished. Incipient problems are caught before they can turn into casualties. Major upgrades can be introduced without significantly disrupting power generation. Such renewable plants can operate indefinitely; but, if a ThorCon is decommissioned, the process is little more than pulling out but not replacing all the replaceable parts.

Cheaper than Coal

ThorCon requires fewer resources than a coal plant. Assuming efficient, evidence-based regulation, ThorCon can produce reliable, carbon-free, electricity at between 3 and 5 cents per kWh depending on scale.

And most importantly, for the case of rapid deployment

The photo on the right is a shot of the Hyundai shipyard in Ulsan, Korea. This single yard can turn 3 million tons of steel plate into over 100 large and complex ships in a year. On average, these ships require about the same amount of steel as a 1 GWe ThorCon. The ThorCon structure is far simpler and much more repetitive. **In short, a single ThorCon yard the size of Hyundai minus the massive building docks could turn out one hundred 1 GW ThorCons annually.**

One of the knocks against nuclear is the plants cannot be deployed in time to make any real dent in coal nor CO2 emissions. For ThorCon, this is not the case. The combination of lower resource requirements and shipyard productivity means that ThorCons can be deployed ***more*** *rapidly than coal plants. It's simply a matter of our deciding to take advantage of this capability.*

The designs of ThorCon and Terrestrial and Copenhagen Atomics (www.copenhagenatomics.com) are made with this kind of fabrication and deployability in mind. A design philosophy which is as old as the Model T Ford. But has been proven to work excellently.

Transatomic Power - the TAP Reactor
Excerpted from www.transatomicpower.com

The nuclear industry of the 1950s was defined by an inexhaustible optimism and rigorous scientific thinking. Anything was possible, and nuclear energy promised to power the world. Revolutionary designs were prolific. Today, however, this technological diversity has been narrowed, and the industry has become locked into one design: the light water reactor. We're challenging this strategy and have returned to the beginning to explore another path, and another design – the molten salt reactor. This simple reactor design, updated with modern technology and materials, has the potential to revolutionize the nuclear industry.

A known drawback of nuclear power is the creation of waste. Some call it a problem: We call it an opportunity. The waste from conventional nuclear reactors can be used as the fuel for our reactors. Light water reactors consume only about 4% of the energy in their uranium fuel, which means that their spent fuel rods contain vast amounts of untapped energy and remain radioactive for hundreds of thousands of years. Our reactors can use this waste, generating enormous amounts of electricity.

The following is an excerpt from their latest white paper.[77]

TAP has greatly improved the molten salt concept while retaining its significant safety benefits. The TAP MSR's primary technical innovations over previous MSR designs are to introduce a zirconium hydride moderator and to use a LiF-based fuel salt. During operation, the fuel in the salt is primarily uranium. Together, these components generate a neutron spectrum that allows the reactor to use either fresh low-enriched uranium (LEU) fuel or the entire actinide component of spent nuclear fuel (SNF). Previous molten salt reactors such as the ORNL Molten Salt Reactor Experiment (MSRE) relied on high-enriched Uranium, with enrichments up to 93% U-235.

Enrichments that high would raise proliferation concerns if used in commercial nuclear power plants.

Transatomic Power's design also enables high burnups – more than twice those of existing LWRs – over long time periods. The reactor can, therefore, run for decades and slowly consume both the actinide waste in its initial fuel load and the actinides that are continuously generated from power operation. Furthermore, our neutron spectrum remains primarily in the thermal range used by existing commercial reactors. We, therefore, avoid the more severe radiation damage effects confronting fast reactor designers, as thermal neutrons do comparatively less damage to structural materials.

The TAP reactor is being developed by a team of bright young scientists who have had their education at the Massachusetts Institute of Technology / MIT.

As you can see, the much-used argument of nuclear waste is actually an argument in favor of nuclear energy and innovation. In fact, we have about 260,000 tons of spent fuel in the world, and this spent fuel—or waste as some like to call it—can be used to produce so much energy that humanity doesn't need to do anything else for the next 70 years. And this is using fuel that is already extracted and above ground. It is highly improbable that we will be building TAP reactors exclusively from the moment they will become available for the market, so we may, with a high degree of certainty, claim that there's enough fuel for TAP reactors for centuries to come.

FliBe Energy - the LFTR

The LFTR is the most advanced iteration of the MSRe. It is being designed by Kirk Sorensen an ex-NASA Space Engineer. He has brought this technology back from its long slumber and has started to work on building a breeder design which is specifically geared towards utilizing thorium as a fuel. Thorium is a fertile element, which means that if you hit it with a neutron it eventually, with a couple of intermediate phases, transmutes into a fissile uranium233 atom, which can be used to create energy in a fission process. Sorensen's design is a dual-fluid reactor which incorporates a closed chemical separation system in order to keep both fluids at optimum performance levels. The LFTR might be the hardest design to realize given the current legislative situation. It could be built in such a way that isotopes could be extracted from it which could be used for proliferation. However, it can also be designed specifically to keep people from doing this. If it is designed appropriately, no-one could ever get any fissile material out of the reactor or its ancillary systems. Besides, this reactor would burn all fissile material. Designing it to be proliferation resistant is paramount. The LFTR is the most promising MSR technology on the table, for it offers virtually unlimited resource flexibility and may help us progress into the thorium epoch indefinitely.

Another fluid fuelled reactor, the MCFR (molten chloride fast reactor)

Terrapower is another forthcoming startup, owned by Bill Gates, which is attempting to create a different reactor design that can run on *nuclear waste*. Here's an excerpt from their website. (www.terrapower.com)

Creating a Safe, Secure Energy Source

Conventional reactors capture only about 1 percent of the energy potential of their fuel. The traveling wave reactor (TWR) represents a new class of nuclear reactor. It is a near-term deployable, truly sustainable, globally scalable energy solution. Unlike the existing fleet of nuclear reactors, the TWR burns fuel made from depleted uranium. This substance is currently a waste byproduct of the enrichment process. The TWR's unique design gradually converts this material through a nuclear reaction without removing the fuel from the reactor's core. The TWR can sustain this process indefinitely, generating heat and producing electricity.

The TWR offers additional benefits over today's light water reactor (LWR) designs:

> ➢ *Provides up to a 50-fold gain in fuel efficiency, which means less fuel producing more electricity. Increased fuel efficiency also means less waste at the end of the reactor's life.*

> *Eliminates the need for reprocessing and significantly reduces and potentially eliminates the long-term need for enrichment plants. This reduces proliferation concerns and lowers the cost of the nuclear energy process.*

> *Directly converts depleted uranium to usable fuel as it operates. As a result, this inexpensive but energy-rich fuel source could provide a global electricity supply that is, for all practical purposes, inexhaustible.*

> *TerraPower is committed to the near-term deployment of TWR technology. We aim to achieve startup of a 600 megawatt-electric prototype in the mid-2020s, followed by global commercial deployment.*

Why are these designs important?

Just as the SunPower E20 435 Watt has been chosen to try to determine the feasibility of the 100%WWS Roadmap, I have selected the AP1000, IMSR, and ThorCon MSR as reactor designs that provide reasons for optimism if build speeds are to be determining factors. The LFTR and the TAP reactor are included to demonstrate that nuclear waste is not a problem, and that it provides opportunities that can be used within a decade.

What about the claim that nuclear cannot ramp up as quickly as possible?

Let's consider Nuclear's past and determine whether previously attained build rates could be augmented or not. Whereas it is true that the steps undertaken before a facility is actually built takes longer in the case of nuclear compared to wind and solar, it does not negate the possibility of high deployment figures for civilian nuclear energy. The extensive planning phase for nuclear reactors is often raised as a showstopper, but is this a valid argument?

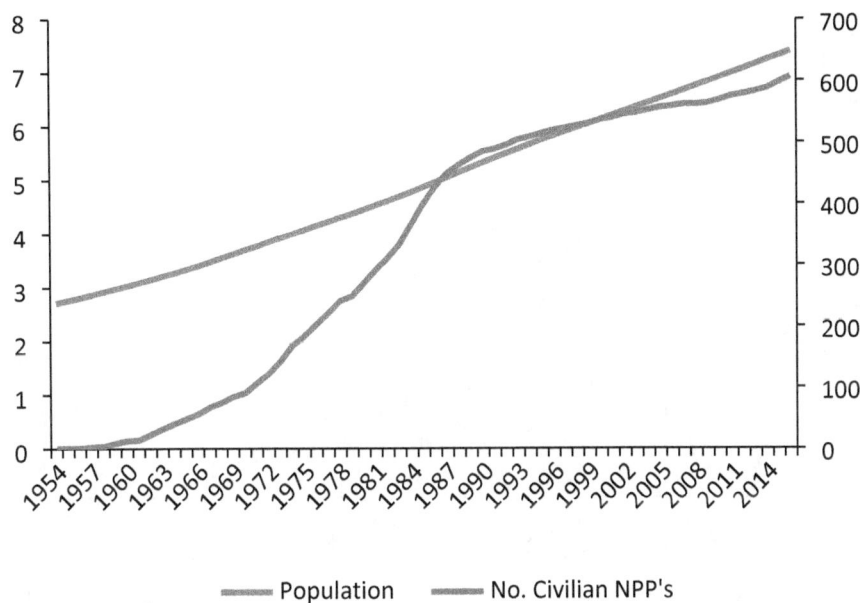

Population — No. Civilian NPP's

Population figures are in billions

The primary axis reveals the linear growth of the world's population since 1954 and the year when we connected the first civilian nuclear power plant to the grid. The secondary axis, displays the number of civilian nuclear power plants connected to the grid in the respective years.

The 1990s and 2000s saw civilian nuclear power expansion decrease due to fear caused by TMI and Chernobyl. However, the red line began to rise again due to regained confidence in nuclear energy, the promise Generation III+ designs, and the willingness of countries to invest in new civilian nuclear energy.

If we plot the annual additions of finished nuclear power plants to the grid we get the following graph.

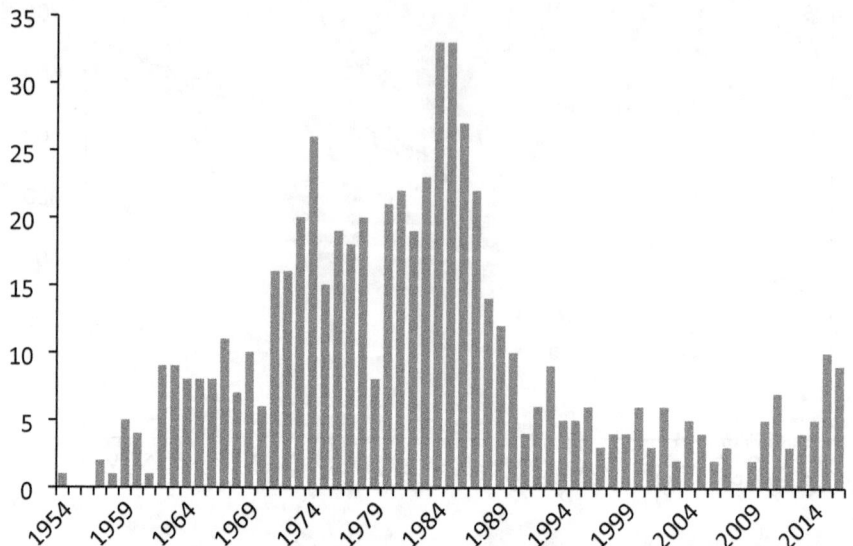

As you can see, the highest number of nuclear power plants were added to the grid in the mid-1980's. If we add a secondary axis so we can plot the number of people per power plant added to the grid, we get the following picture.

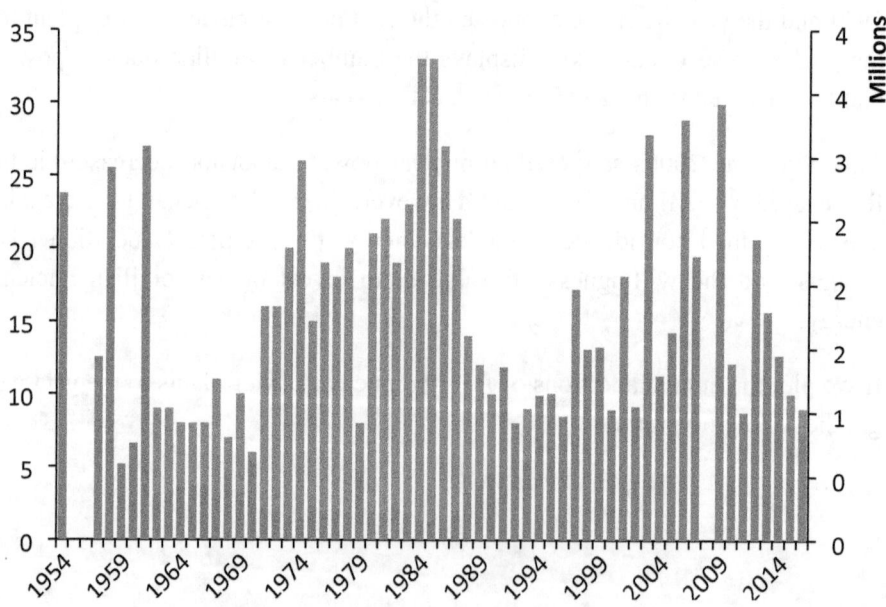

This graph reveals that we were capable of adding one civilian nuclear power plant per 144 million people in 1982, which might seem like a trivial figure, but bear with me. We currently have about 7.4 billion people on Earth. How many reactors would we be able to add if this metric were used to determine feasibility?

$$\frac{7{,}400{,}000{,}000}{144{,}000{,}000} = 51.4 \; Civilian \; Nuclear \; Reactors$$

Starting to build 51 reactors isn't that difficult, in fact, we've already started to build 37 reactors in 1967 and 1969, and our capabilities have grown significantly since then.

The important thing to note is that we are talking about 51 generation II reactors, as they were the prevalent technology back then. Generation III+ and generation IV reactors are easier to build because they are less complicated, and have smaller footprints. Also, note that most of these power plants have been constructed in countries which are technologically advanced. This metric could be improved as overall levels of education and scientific literacy increase all over the planet—which is not a given, but something we should aspire towards.

If all of these reactors were 1 GW units, they would generate about 405 TWh each year. By this measure, it would take us 617 years to decarbonize all energy (if 250,000 TWh), and 86 years to decarbonize all electricity (~35,000 TWh, page 69). Nowhere in the wind and solar industry are solutions of this magnitude feasible—like the possibility of constructing nuclear power plants that were manufactured in a factory. Never before have we used automated assembly line processes to fabricate reactor cores and the modules we need to build a complete nuclear power plant, but it is possible, and it will revolutionize the way we do things.

ThorCon Power, which is run by experienced professionals with an industrial background in large ship building, has quantified its hypothetical capabilities.

It is also important to note that building a nuclear power plant isn't much more difficult than building a coal-fired power plant, or a gas-fired power plant, and we can build hundreds of them worldwide in a year. Did you know that a nuclear reactor takes 6.7 years to get built on average, but that it can be done in as little as 1.8 years. Japan used to be by far the best country to build a nuclear reactor as it only takes 4.1 years to complete a reactor on average. It is probable that standardized, modular reactors can cut these times significantly. The standardization of reactor designs will also help greatly in streamlining the processes that come before the start of the build. It is safe to assume that reactors can be delivered within a timeframe of 5 and 10 years from initial planning until grid connection. While it is true that there are projects which take too long to complete, these are hardly the norm.[78]

If the EIA is right, and their 2012-2040 electricity prediction (Page 56) is accurate, and if we want to use nuclear power to offset gas and coal, we would need to do the following:

First we plot the graph from year to year based on the EIA prediction for electricity. Note that the primary axes are in Thousand TWh.

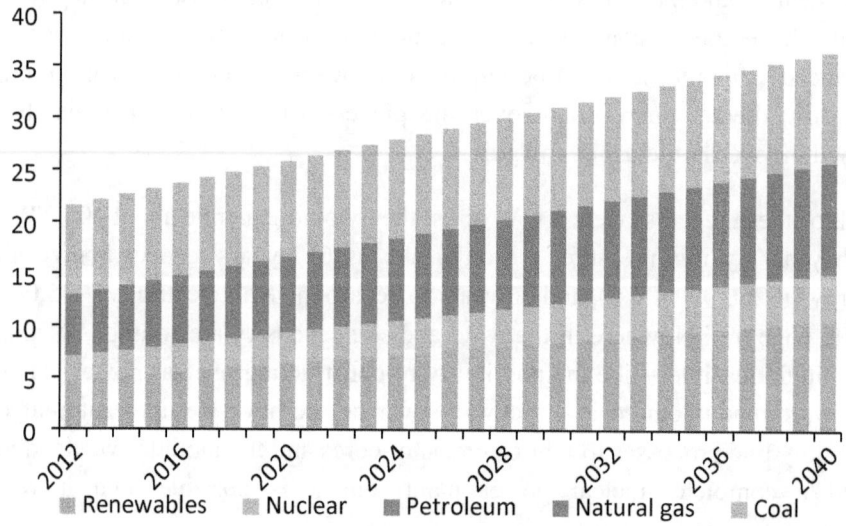

Figures in Thousand TWh

Now we plot the same graph while accounting for a linear decline of the fossil fueled electricity technologies and replacing them with nuclear.

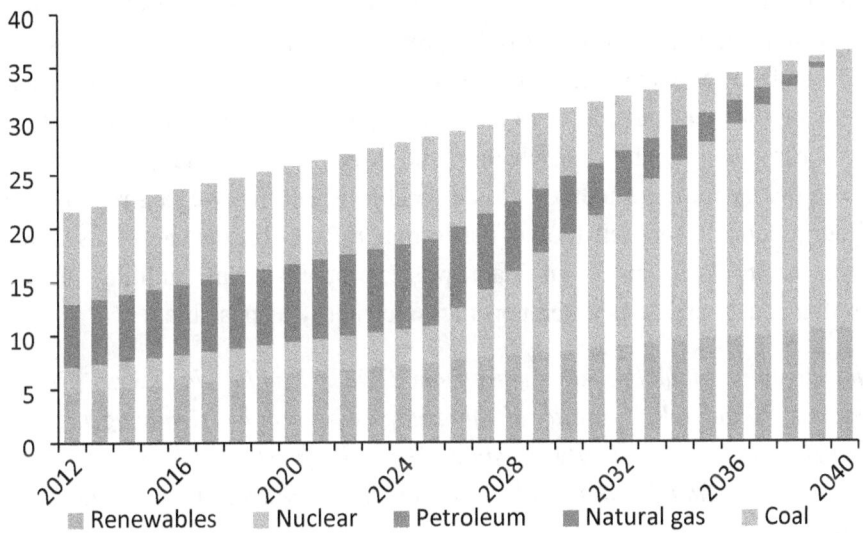

Figures in Thousand TWh

With this scenario, we would achieve decarbonization of electricity within 28 years while accounting for some lag in nuclear developments and the need to offset the effects of "the nuclear cliff". The connection rate before 2025 is as the EIA has predicted, but after 2025, nuclear generation capacities have to be increased with 1500 TWh per year. How many 1 GW reactors is this?

$$\left(\left(\frac{1500\ TWh \times 1000}{90}\right) \times 100\right) \times 8766\ hours = 190\ GW$$

Given that startups like ThorCon show that 100GW per fabrication yard might be possible, this seems feasible. We could decarbonize electricity before 2040 by deploying generation IV MSRs, PBRs, MCFRs, Sodium Breeders, and even Generation III+ AP1000s and APR1400s and a great host of Russian VVER type reactors. However, the challenge is far greater than "just decarbonizing electricity", which is why I call for increased R&D on HDR (Hot Dry Rock) Geothermal as well. We need to augment the deployment rate of high-density low-carbon energy sources, which is something Jacobson and I actually agree on.

Once all these modern reactor designs become viable in a commercial sense, we will truly enter the "Thorium Age", an age in which nuclear fuel is basically limitless and will last for the duration of the planet's own lifespan. It is also important to note that we are probably close to being able to extract uranium from Seawater at parity, which means that it will become just as cheap, or cheaper than contemporary mining processes. Also, extraction from the oceans is probably far cleaner and leaves less waste products than contemporary mining methods. It is a good prospect that we might be able to utilize practically all fertile and fissile materials to create energy as it will enable us to keep any significant waste problem from building up and simultaneously create as much non-invasive energy as possible.

"New technological breakthroughs from DOE's Pacific Northwest (PNNL) and Oak Ridge (ORNL) national laboratories have made removing uranium from seawater economically possible. The only question is – when will the source of uranium for our nuclear power plants change from mined ore to seawater extraction?

Researchers at PNNL exposed this special uranium-sorbing fiber, developed at ORNL, to Pseudomonas fluorescents' and used the Advanced Photon Source at Argonne National Laboratory to create a 3-D X-ray microtomograph to determine microstructure and the effects of interactions with organisms and seawater. Courtesy of PNNL

Nuclear fuel made with uranium extracted from seawater makes nuclear power completely renewable. It's not just that the 4 billion tons of uranium in seawater now would fuel a thousand 1000-MW nuclear power plants for a 100,000 years. It's that uranium extracted from seawater is replenished continuously, so nuclear becomes as endless as solar, hydro and wind."
—James Conca, ANS Nuclear Cafe, October 3rd, 2016 [79]

Designing and building a reactor and its supporting infrastructure takes time, and it will not be easy. But, we have great capabilities and, with our technological ingenuity, have built many large and complex projects. Building these reactors will be much the same as building airplanes, ships or trains, which we do by the thousands. Generation IV reactor deployment will no longer be limited by large foundry capabilities—which are currently needed to forge the pressure vessels required by contemporary pressurized water reactors.

The challenge remains amazingly large. But with nuclear technology at the table, meeting the challenge at least enters the realm of the possible. It helps us think in terms of the world we have today and the world we are likely to have in 2050, finding credible solutions to real problems, rather than dreaming in the invented WWS scenario that tailors a problem to preferred solutions.

Speculations about a future of non-carbon energy

*Speculation — **noun** — the forming of a theory or conjecture without firm evidence.*

Note the use of the word theory, which I am not using in the scientific sense because one cannot form a scientific theory without sufficient evidence. The reason why the Theory of Evolution is called a theory is because it has been tested exhaustively, and the evidence confirms its predictions every time.

Thinking about future energy mixes involves speculating. The 100%WWS Roadmap has tried to present a considerable body of evidence to support the claims of the lead author. But the fact that some prerequisites for the roadmap are unattainable forces us to question its feasibility. Moreover, the 100%WWS Roadmap is rife with overly optimistic assumptions.

To get a clearer perspective on energy matters, we must re-examine the 100%WWS Roadmap and acknowledge the need for an alternative plan that has the same aim: the elimination of fossil fuel emissions. However, we also need to take a slightly more realistic approach by including fresh water needs (through desalination, increased water management, and waste treatment); carbon capture and permanent sequestration through chemical alteration; Ocean de-acidification practices; efforts to decouple humanity from nature in order to allow the biosphere to recuperate from our negative influence upon it.

My counterargument is this: In order to decarbonize human civilization and achieve this before 2050 or 2075 at the latest, we need a foundation of nuclear energy and augmentation of the energy mix with hydro, geothermal, wind, and solar.

It is also necessary to consider decarbonization pathways, which encounter widely varying levels of political commitment. We really do face an "Apollo

Program" challenge, but, pathways that require what Bill Mckibben calls "World War Mobilization", which focuses on decarbonization, but pays little attention to socioeconomic issues, will likely be less productive than a more inclusive agenda.

let's examine a possible scenario, and you can decide if it is feasible or not. I've created an acronym for the following energy mix: nuclear 60%, geothermal 25%, hydro 5%, wind 5%, solar 5% - **NGHWS**.

The end of all fossil fueled processes by 2050.

It is highly unlikely that an energy mix that relies on wind and solar alone will suffice. Instead, we must create an energy mix composed of nuclear, geothermal, hydro, wind, and solar. Besides creating the generation capacity necessary to satisfy 2050s demand, we will need to consider the cumulative upkeep of the renewable part of this mix for at least 50 years. We are already stretching the Earth's resources, so we have to become more efficient.

For this optimistic future to become a reality we have to deploy all of the possible innovations and keep pushing the boundaries of technological and scientific understanding.

Two "innovations" are required to meet this objective: 1. We need to replace coal furnaces with nuclear reactors; 2. we need to start building nuclear reactors on assembly lines. Fortunately, Terrestrial Energy and Thorcon Power are making serious progress, with Terrestrial Energy holding pole position due to their extensive involvement with the Canadian and US Governments and their respective agencies and research institutes.

I'll stick to this hypothetical NGHWS mix: Nuclear 60%, Geothermal 25%, Hydro 5%, Wind 5%, Solar 5% while accounting for startup times and/or lag for the nuclear innovations to become available.

First, real additions will start at 2025. The graph will begin close to 10,000 TWh (for the NGHWS portion), whereby 2025 roughly 195,000 TWh is projected by the EIA, and this gives us an idea of the gap between low-carbon and high-carbon energy sources.

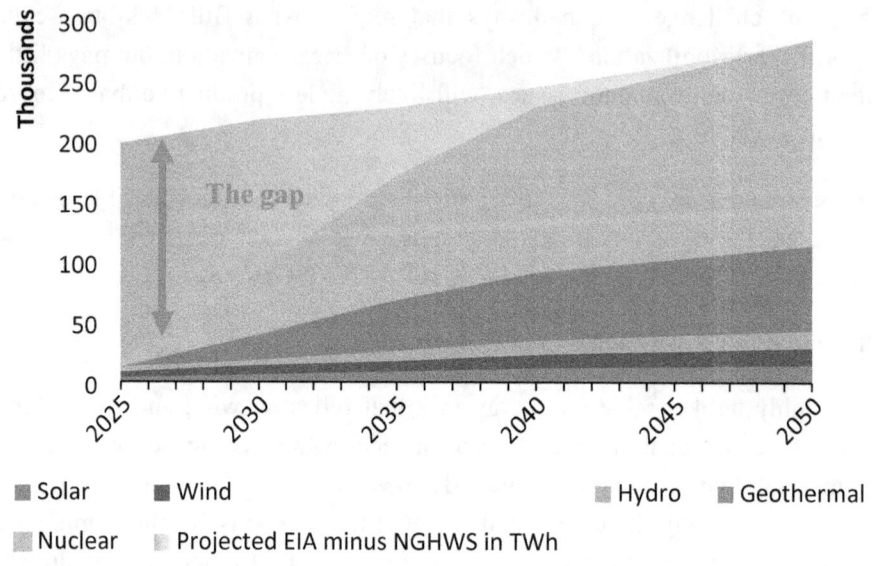

Figures in TWh

All growth from 2050 onward should be exclusively NGHWS because a rapid increase is required in order to stem carbon emissions as quickly as possible. This means that the economic viability of all these technologies must be improved significantly. The following graph shows the capacities required for each technology to get where we want to be in 2050.

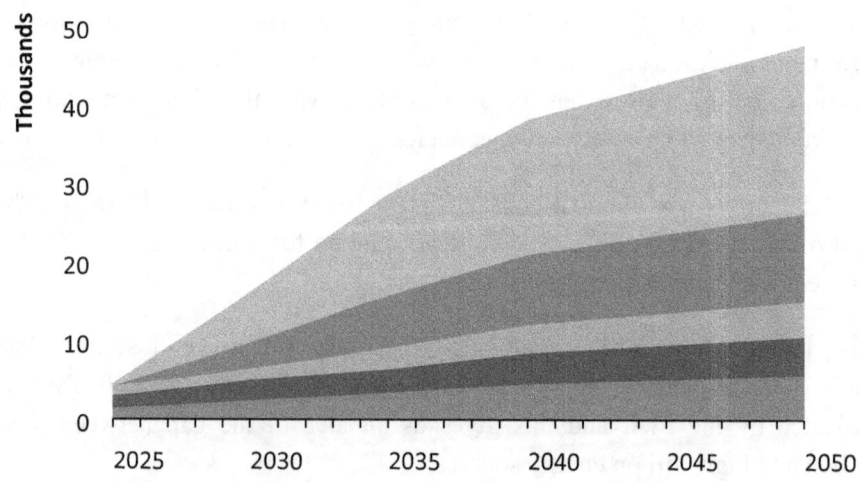

Figures in GW
Colors correspond to the previous graph

The following graph, which is based on capacity rather than on annual generation, reveals a significant change due to the large differences in capacity factor between the different technologies. Suppose we want to generate 250,000 TWh by employing one technology exclusively, we would get the following picture.

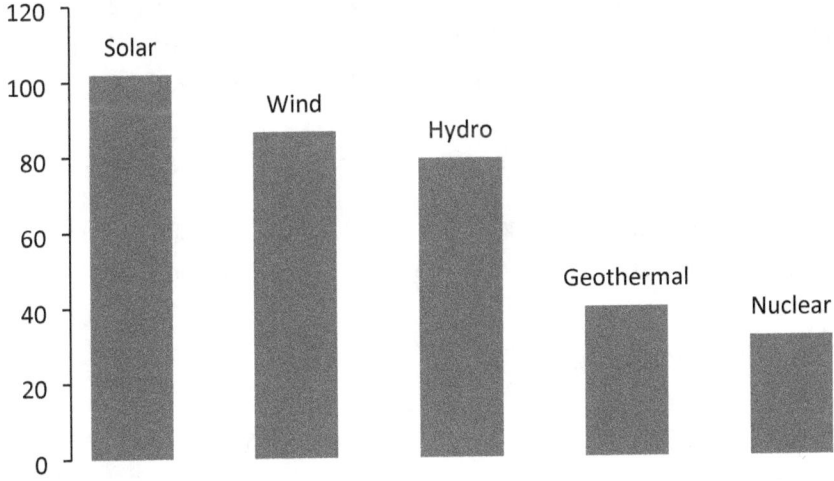

Figures in TW

A nuclear plant with a capacity of roughly 32 TW can generate 250,000 TWh per year, but solar requires a capacity of about 102 TW to generate the same amount of energy, a telling discrepancy that should make us re-examine the value of solar energy (and wind, and hydro).

If we extrapolate the growth projected by the EIA we find that we may expect total energy production required to be roughly 282,000 TWh (page 69) by the year 2050, and if we haven't managed to curtail growth by then it may even grow to 615,000 TWh by the year 2100. I will continue to work with the 282,000 TWh figure, just to show you how steep the challenge eventually may be. Note that future energy production may be lower or higher, it is uncertain at this point. However, the predictions by the EIA should be taken seriously.

Technology	Name-Plate Capacity in GW	Capacity Factor	Capacity per unit in MW	Individual Units needed	Total annual yield in TWh
Nuclear	21,468	90	800	26 835	169,371
Geothermal	11,228	71.7	150	74 854	70,571
Hydropower	4485	35.9	x	x	14,114
Wind	4879	32.5	5	9.76E+05	14,114
Solar	5630	28.6	1	5.63E+06	14,114
Total	47,690				282,285

Total 5MW wind Turbines	1 Million
Total 435 Watt Panels	13 Billion

Let's recall the roadmap figures:

Total 5MW wind Turbines	2 Million
Total 435 Watt Panels	75 Billion

As you can see, my "optimum" scenario cuts Jacobson's demand for 5MW wind turbines in half, and the solar requirements by a factor of 6.25. Such rapid and massive expansion of solar and wind is highly problematic, particularly in regard to materials used, chemistry required and the environmental impact, including bird and bat death prints. However, I do cede that wind turbines have a small role to play in remote, gridless areas. I would rather have windmills and solar panels form 1% of the total energy mix with HDR geothermal and nuclear energy dominant, the reasons by now should have been well advertized.

Now let's consider a final energy mix per 2050 based on five technologies, primarily nuclear and HDR geothermal:

Technology	Name-Plate Capacity in GW	Capacity Factor	Capacity per unit in MW	Individual Units needed	Total annual yield in TWh
Nuclear	24,957	90	800	31,196	196,894
Geothermal	10,442	71.7	150	69,614	65,631
Hydropower	4,485	35.9	x	x	14,114
Wind	991	32.5	5	1.98E+05	2,823
Solar	1,126	28.6	1	1.13E+06	2,823
Total	42,001				282,285

Total 5MW wind Turbines	199 Thousand
Total 435 Watt Panels	2.6 Billion

In this scenario, the number of wind turbines and PV panels required have diminished drastically, but the deployment figures for these technologies will probably be higher in the future. However, let's be cautious about misallocating resources because we humans often over commit, which inevitably leads to new problems.

This final scenario shows that we need to build roughly 100,000 nuclear reactors and HDR geothermal plants. The build-rate required for nuclear is a staggering 1250 units per year, and the build-rate for HDR geothermal plants is even higher—2800 units per year. Is this feasible? Yes. is it necessary? Absolutely! Do I think this is realistic? Absolutely not.

Even the technology with the best rate of energy returned on materials invested must be deployed without decimating natural resources. I would welcome a future of plentiful energy for everybody, regardless of the cost, but we cannot ignore the cost. As long as capitalism and democracy dominate our decision-making processes, we will have to make do with what we have. And when special interests and demagogy dictate our energy policies, we will be mired in indecision. In the meantime, the world placidly zooms around the sun, and the climate keeps a-changing for the worse, regardless of whether

President Elect Donald Trump (November 15, 2016) and his administration believe it or not...

We have not yet compared the amount of copper required by nuclear power and renewables, so let's consider it now.

Nuclear energy (in the US in 1974) required roughly 0.73 Kilograms of copper and brass per Kilowatt of capacity,[80] which means that a 1000 MW nuclear plant would contain about 730,000 kilos (730 Metric Tons) of copper.

In contrast, 1000 MW worth of Vestas 2 MW wind turbines[81] contain 3,320,000 Kilos (3320 Metric Tons) of copper, and when we compare total materials required per kW even a 1974 generation II Reactor beats a wind turbine with 216 Kilos/kW for nuclear against 542 Kilos/kW for wind. Nuclear power is at least twice as conservative in materials used per kW capacity built, without accounting for capacity factor.

If we include capacity factors, the figures look like this: 216 Kilos/0,9kW for nuclear and 542 Kilos/0.325kW for wind or 240 Kilos/kW(nuc.) and 1668 Kilos/kW (wind).

Wind power simply cannot compete with nuclear power on these most important of all metrics. However, we will keep working on the original premise of 216 and 542 kilos/kW because we have accounted for capacity factor in the respective growth scenarios required. Again, we will push the start-up towards 2025 and assume that wind and solar keep growing at their current rates until 2025.

How much copper is required to build a generation capacity of 100,000, 150,000, 200,000, 250,000 TWh of nuclear energy? And let's contrast these values with those for wind and solar, working off the total 5.5 Metric Tons/MW figure, which is fair given the fact that wind and solar need additional copper resources for transmission. First, the nuclear chart.

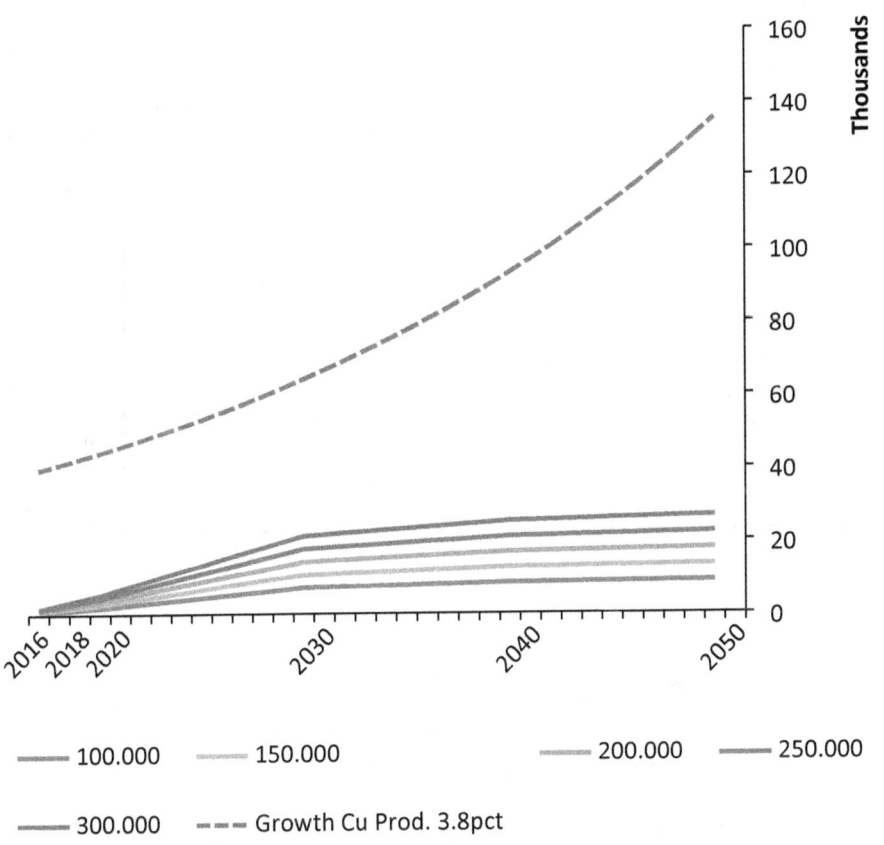

Figures in Thousand Metric Tons.

The growth rate of Copper in this graph is exactly the same as the one we've seen earlier (3.8pct). As you can see, nuclear barely reaches the 30,000 thousand metric tons figure, it doesn't even scratch total Copper production.

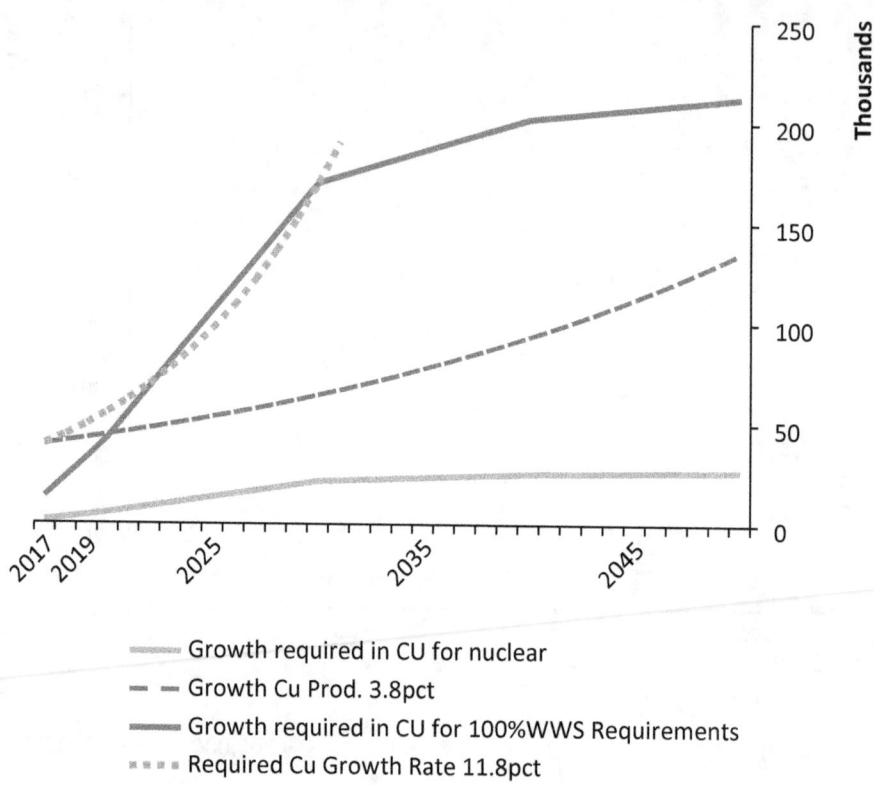

- Growth required in CU for nuclear
- – – Growth Cu Prod. 3.8pct
- Growth required in CU for 100%WWS Requirements
- ▪ ▪ ▪ ▪ Required Cu Growth Rate 11.8pct

Figures in Thousand Metric Tons.

Here we have contrasted the highest possible copper requirement for nuclear (300,000 TWh) with the lowest possible copper requirement for wind and solar (100,000 TWh). Note the blue dotted line, which is the required growth rate to keep up with the 100,000 TWh target for WWS. It is a growth rate of Cu production of roughly 11.8%. WWS technologies cannot compete with nuclear in terms of material efficiency. Not only does nuclear outperform the renewables, even with uranium extraction it doesn't even come near the extraction required for the 100%WWS Roadmap to succeed. You've seen the additional context in which wind requires at least twice the amount of materials per capacity, without accounting for capacity factor. I trust that you will, by now, understand what that would look like if put in a graph. It is also important to note that we've not accounted for lifespan of the different technologies, where nuclear power plants life spans range from 50 to 80 years, wind turbines and PV panels only reach 20 to 25 years, which is an additional consideration, because the amount of materials needed doubles much quicker when compared to nuclear energy. The copper matter is a complex one as demand and availability fluctuate continuously in a highly volatile market. When you try to look for long-term prognoses on copper availability, you won't find a lot of information. It simply is that hard to prognosticate. The richness of the ores is an uncertain factor in which it is more probable that we will have to use increasingly less rich ores, and this will the drive energy required, and thus price, to extract it, upward.

Keep in mind that we have been using figures associated with 1974 generation II Reactors. It is not known to me whether generation III+ reactors use more or less copper than their older generation II counterparts. However, what does Westinghouse have to say about this? Compared to their own generation II Reactor, their generation III+ reactor (The AP1000) has 50% fewer valves, 35% fewer pumps, 80% less pipe, 45% less seismic building volume, and 85% less cable[82]

As you can see, by their own calculations their generation III+ nuclear power plant has significantly fewer material requirements than the "contemporary" generation II reactors that dominate the nuclear landscape today. All you need to do is take a glance at a top-down schematic of generation II and generation III+ power plants to spot the difference.

Simplification - Smaller Footprint

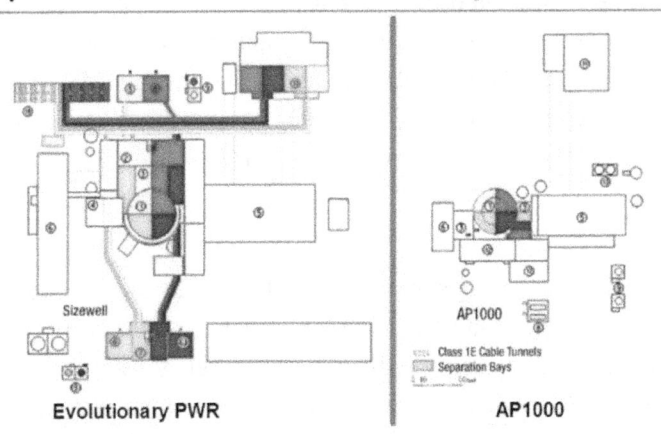

Image credit: Westinghouse,
http://www.iaea.org/inis/collection/NCLCollectionStore/_Public/42/026/42026956.pdf

We're on the verge of a new dawn in civilian nuclear energy in which smaller, more efficient, reactors will be deployed faster than ever before. New designs will provide a solution to the "materials problem" which was the main focus of this book. If we want to build a sustainable and optimistic future, we have to focus on the availability of materials and the energy it will take to extract them, and make them suitable for production and manufacture. Even though our contemporary energy context is mainly driven by monetary economics, I will assert that the economics of the future isn't about money or fuel, it will revolve around raw materials.

"I am sorry, we wanted to save the biosphere, but we didn't have the money to do it" is different than *"I am sorry, we wanted to save the biosphere, but we didn't choose the right technology to do it"* – but no better from future generations' point of view. It is, the likely assessment of the outcome if we follow Jacobson's example.

We have to move with *"all deliberate speed."*

All deliberate speed is a term which has come to my attention by a video called *"Young People's Burden"* which has been uploaded by Sophie Kivlehan and her grandfather James Hansen, the well-known scientist of the Earth Institute at Columbia University. This term made me think about the nature of the word deliberate. If we move forward we should do so by design. We know that we have to reduce greenhouse emissions and this should be the main focus while simultaneously increasing energy generation for our growing population. As has been shown, we also need to start working on reducing atmospheric carbon levels and restoring the hydrological cycle, which means that we have to engage in geoengineering. The *"deliberate"* part hides within the details of how you move forward.

How we get there?

> ➤ Each coal-fired power station should be closed as soon as possible (and no new ones may be built) while being replaced by nuclear or geothermal power plants with at least equal generation capacities. The use of high-temperature nuclear generators may enable the preservation of the balance-of-plant of the coal stations, making this quicker and cheaper.

- Each gas-fired power station should be closed as soon as possible (and no new ones may be built) while being replaced by nuclear or geothermal power plants with at least equal generation capacities.
- Small communities can start their path to decarbonization by utilizing small-scale wind, solar, and hydropower. However, Geothermal should be the preferred technology.
- Transportation has to be decarbonized as fast as reasonably possible, without exacting a heavy toll on natural resources, and without putting unreasonable pressure on the people that live in the vicinity of the mining and purification and production facilities.
- Agriculture needs to be revolutionized and demand for crops and meats with high greenhouse gas emissions should be decreased.

Most of these issues are ones of market economics but are also in the hands of policy and lawmakers. "*All deliberate speed*" can only be attained if there's a regulatory framework to make decarbonization happen, which means that we have to force the governments of our planet to acknowledge the seriousness of our worldly situation, and take responsibility by acting upon it passing deliberate legislation that pushes the implementation of effective solutions to the climate change problem with *all deliberate speed*. Agreements, like those made in Paris at COP21, are being heralded by participating governments as being revolutionary. However, they are grossly inadequate in addressing our current situation. What needs to be done is this:

- Implement Legislation based on Carbon fee and dividend, which is a natural way of incentivising low-carbon technologies over high-carbon technologies.
- Boost R&D on nuclear, geothermal, wind, solar, hydro, agriculture, and transportation technologies.
- Enact stringent license renewal schemes for coal and gas-fired power plants.
- Ask and encourage the utility companies to implement nuclear, geothermal or in small-scale cases wind, solar, hydro.

The energy crisis that is enveloping humanity, however, calls for desperate measures. Our world leaders will, at some point, be forced to act or admit their inability (or unwillingness) to act. For now, the focus is on replacing coal, mostly with natural gas, but also with some other technologies. Even though many believe these other technologies to be wind and solar, I think I have presented evidence to the contrary well enough for the reader to decide whether this is feasible or not. I think that 100%WWS or NGHWS have big challenges ahead. I believe that we can decarbonize energy production, but I don't think that it is going to happen by 2050; it is far more reasonable to expect full decarbonization or at least an exit from fossil fuels to happen by 2075 or 2100. In the meanwhile, we should have figured out how to keep the pyramid of marine life from collapsing, and to have built an extensive water desalination and management network. One thing is certain, it won't be an exclusive 100%WWS energy mix that will power civilization for decades to come, it will be a mix of nuclear and geothermal augmented with hydro, wind, solar, and a couple of other technologies that might not yet have been developed, or matured. Until we reach decarbonization, we have to acknowledge the fact that coal, gas, and oil will be adding CO_2 to an already substantial carbon debt.

Conclusion

What is the measure of success?

Some would say that implementing a future that is 100% renewable would be a success. In fact, it is often said and written that the 100% renewable future is the aim. Or that our future should be green, and renewable. But this isn't the true measure of success. I would like to take it one step further. Success should be determined by how quickly and efficiently we could end the combustion economy, and address additional issues such as ocean acidification and decreasing fresh water availability. Where the previously mentioned "efficiency" means the most provident use of resources, as we are responsible for the Earth that subsequent generations of humans will inherit. I suppose that both Jacobson's and my objectives are to dive deep below IPCC's RCP2.6 scenario. We have to stop CO_2 emissions in order to keep the atmospheric and oceanic concentrations of CO_2 manageable. The aim should be negative CO_2 emissions from anthropogenic processes. We should reduce our own carbon footprint as much as possible. However, we should not drop CO_2 levels below certain levels, because we might inadvertently trigger another unwanted climactic response. CO_2 levels in the atmosphere have to be within a safe margin that will enable life on Earth to thrive.

Decoupling from nature is absolutely necessary in order to help the biosphere remain healthy, and will also serve to help terrestrial species survive a changing world. To reach these goals we have to do the following:

- Change our consumption patterns, become more efficient and resourceful. For our natural resources are finite and the destruction of nature needs to be curtailed significantly;
- We have to find a way to end the fossil combustion economy as **quickly** as possible and progress into an electrified age in which

combustion is only based on clean synthesized fuels, but kept to an absolute minimum.

What serious ecological problems may we expect when we fail to get serious about decarbonising civilization:

- Faster sea-level rise, through collapsing ice sheets and diminishing snow packs and glaciers; affecting roughly 1 billion humans directly, and practically all of the economic hotspots of the world, perhaps even causing massive financial instability;
- Stronger storms and storm surges in heavily populated areas;
- A spread of droughts, eventually leading to desertification;
- An increase in oceanic dead zones;
- A possible mass maritime extinction due to acidification from unmitigated carbon dioxide emissions;
- An increase in pandemics and epidemics (Zika, Ebola, Malaria and the suchlike).

It is also important to take a look at man-made climate change and what it could mean for you, and your family:

- A family member may die from a heat wave, or from a disease caused by the combustion economy;
- The price of your food will go up, it is not a matter of IF, but WHEN;
- Some forms of food might become scarce or might even disappear;
- You may have to run your air conditioner more;
- A migrant crisis caused by man-made climate change might affect policies and taxes in your country. (please note that I don't demonize migrants because I am a humanist);
- If you live on a shore, or close to a river, the likelihood of damaging floods and surges increases;
- If you live in an area prone to tropical cyclones, tornados and the suchlike, these will become stronger and more damaging;
- Water may become rationed. Look at California where watering your lawn has become a policy issue;
- Sustained drought might render the area where you live uninhabitable;

- Sustained drought might also cause severe forest fires, threatening your home, or the home of someone you know;
- Think about the problems of New Orleans and project them to many low-lying coastal cities. If you live in one of these cities you may or may not see your house become flooded, but perhaps those of the people living down at the shoreline;
- Extreme weather such as storms, heavy rain, blizzards, and droughts

Likelihood is the most important word to add here. The likelihood of these things happening increases with man-made climate change getting worse as we fail to curtail greenhouse gas emissions. Another keyword is collapse. We will see natural and human systems collapse if we fail.

This roadmap is a model with predictive capabilities in energy generation and consumption. But it fails on multiple levels, and deserves, therefore, to be falsified. A new model to counter his idea, and its momentum, is required. The critical shortcomings of the 100%WWS roadmap summarized:

- Between 60,000 and 110,000 TWh are unaccounted for;
- the 100% WWS roadmap does not account for geoengineering;
- Despite decades of heavy investments, renewable growth rates still are insufficient to replace fossil fuels;
- Material requirements and their impact on nature have not been weighed correctly;
- Timelines that demand full decarbonization before 2050 are feasible, however, the practical realization of these timelines is highly unlikely. Even when we choose a different path which is lined with nuclear and HDR geothermal.

Counter arguments:

- Decarbonizing electricity before 2050 using a mix of nuclear, geothermal, hydro, wind and solar power is feasible and likely, provided the presence of commitment;
- Nuclear energy outperforms wind and solar in terms of generation capacity per unit of material input, especially when accounting for capacity factor, and even more so when accounting for capacity

factor and lifecycle. The end result is a factor of superiority in generation capacity per unit of materials invested of >10x;
- Decarbonizing all energy before 2050 using a mix of nuclear, geothermal, hydro, wind and solar is feasible, however, very unlikely due to limited possibilities in material production rates;
- In order to keep within reasonable bounds of materials extraction, we need to put emphasis on developing modular reactors. This requires an increase in R&D on a multitude of different scientific fields such as nuclear engineering, material sciences, chemical sciences, geology, etc.;
- We can use existing coal-fired power plants and convert them into nuclear power plants by keeping the generation and cooling facilities and installing a nuclear reactor where once was the coal burning part of the power plant;
- R&D on Geothermal needs to be increased in order to develop an alternate, low-materials/high-yield energy source, preferably Hot Dry Rock (HDR) Geothermal;
- There needs to be an emphasis on the cogeneration capabilities of nuclear and geothermal in order to maximize material efficiency and minimize energy wasted.

A certain hubris about the prospects for the Solutions Project seems to cloud public judgment. Rather than accepting its claims on face value, I think it far more prudent to take a breath and examine the project and its underlying ideas as critically as possible. Also, note that the 100%WWS Roadmap has passed the process of peer-review. However, this doesn't automatically mean that it is an accurate depiction of reality, or of the future. The predictive capabilities of the 100%WWS Roadmap are questionable. Although this book is insufficient when it comes to falsifying the roadmap, I hope it has provided some material for fruitful and edifying discussion. I suspect that the paper on the same topic submitted by Caldeira et al will be far more persuasive in academic circles.

In conclusion, *The 100%WWS Roadmap/The Solutions Projects* and its underlying hypotheses do not explain how human civilization can decarbonize; neither does it provide a credible solution to the ever growing problems of anthropogenic oceanic change and diminishing fresh water supplies; additionally, it creates a great burden as it wants to eradicate the entire nuclear industry, which is essential in certain fields of medicine and energy generation. Most importantly, the Solutions Project is gravely inadequate when fighting carbon emissions are concerned, or the serious health issues and deaths that stem from the combustion economy. As such authors and activists should be called upon to seek a new hypothesis that includes nuclear energy as part of a strong decarbonization roadmap.

I call upon the scientific community to falsify the underlying hypotheses of the *100%WWS Roadmap/Solutions Project* in order to disenthrall humanity from the idea that we can save civilization and the biosphere upon which it depends, using mainly and almost exclusively wind and solar power.

Even though this pale blue dot is all we have, this is a big world, without evident borders. Fixing the ills of our age requires a well-coordinated effort which can only be animated by reason and science. I hope that someone, somewhere, will convince world leaders to adopt an all-inclusive non-carbon energy hypothesis, rather than a competing, in essence exclusive, wind and solar energy hypothesis which is the 100%WWS Roadmap—a non-solution. It may be a compelling idea, but that doesn't make it true, or probable.

Appendix

References

My references are [almost] exclusively online because I want you to be able to have access to all the sources I've used. I want to show you that you can find reliable information online. But... Always remain skeptical, and accept evidence only from reputable sources [after careful examination, if possible]. It takes some effort to weed out the nonsense, one of the problems when stepping outside the realm of peer-review.

This does not equal an academic bibliography. However, it is an attempt to show you how to source evidence to support your claims

[1] *Giving tax credits to the wind energy industry is a waste of time and money.*
http://www.usnews.com/opinion/blogs/nancy-pfotenhauer/2014/05/12/even-warren-buffet-admits-wind-energy-is-a-bad-investment

[2] *EIA, Electric Power Monthly, with Data for June 2016*
https://www.eia.gov/electricity/monthly/pdf/epm.pdf

[3] *U.S. Atomic Energy Commission, Civilian Nuclear Power ... A report to the president—1962*
http://energyfromthorium.com/pdf/CivilianNuclearPower.pdf

[4] *EIA, International Energy Outlook 2016, Chapter 1. World energy demand and economic outlook*
http://www.eia.gov/forecasts/ieo/world.cfm

[5] *Clean Energy Manufacturing Analysis Center, Automotive Lithium-Ion Cell Manufacturing: Regional cost structures and supply chain considerations*
http://www.nrel.gov/docs/fy16osti/66086.pdf

[6] *Sequence Omega, Tesla's first Lithium Agreement*
http://sequence-omega.net/2015/08/teslas-first-lithium-agreement/

[7] *ETF Daily News, Tesla might have a big problem finding all the heavy metals it needs*
http://etfdailynews.com/2016/08/14/tesla-might-have-a-big-problem-finding-all-the-heavy-metals-it-needs/

[8] *Clean Energy Manufacturing Analysis Center, Lithium Ion Battery Key Elements — More Than Just Lithium*
http://www.manufacturingcleanenergy.org/blog-20160413.html

[9] *U.S. Geological Survey, Mineral Commodity Summaries, January 2015, Lithium*
http://minerals.usgs.gov/minerals/pubs/commodity/lithium/mcs-2015-lithi.pdf

[10] *Battery University, BU-308 Availability of Lithium*
http://batteryuniversity.com/learn/article/availability_of_lithium

[11] *U.S. Geological Survey, Mineral Commodity Summaries, January 2015, Lithium*
http://minerals.usgs.gov/minerals/pubs/commodity/lithium/mcs-2015-lithi.pdf

[12] *International Journal of ELECTROCHEMICAL SCIENCE, Electrical Efficiency of Electrolytic Hydrogen Production*
http://www.electrochemsci.org/papers/vol7/7043314.pdf

[13] *Wikipedia, Fuel Cells*
https://en.wikipedia.org/wiki/Fuel_cell

[14] *Wikipedia, Ocean Acidification*
https://en.wikipedia.org/wiki/Ocean_acidification

[15] *Albright, R., L. Caldeira, J. Hosfelt, L. Kwiatkowski, J. Maclaren, B. Mason, Y. Nebuchina, A. Ninokawa, J. Pongratz, K. Ricke, T. Rivlin, K. Schneider, M. Sesboüé, K. Shamberger, J. Silverman, K. Wolfe, K. Zhu, & K. Caldeira, 2016: Reversal of ocean acidification enhances net coral reef calcification. Nature, doi:10.1038/nature17155.*
http://globalecology.stanford.edu/labs/caldeiralab/Caldeira_research/Albright_OneTree.html

[16] *YouTube, Dr. Ken Caldeira Talk*
https://www.youtube.com/watch?v=ELDvbMIqs3A

[17] *Kwiatkowski, L., B. Gaylord, T. Hill, J. Hosfelt, K. J. Kroeker, Y. Nebuchina, A. Ninokawa, A. Russell, E. B. Rivest, M. Sesboüé & K. Caldeira, 2016: Nighttime dissolution in a temperate coastal ocean ecosystem increases under acidification.* Scientific Reports, doi: 10.1038/srep22984.
http://globalecology.stanford.edu/labs/caldeiralab/Caldeira_research/Kwiatkowski_PMTidepools.html

[18] *NOAA, National Ocean Service, What is coral bleaching?*
http://oceanservice.noaa.gov/facts/coral_bleach.html

[19] *Mathesius, S., M. Hofmann, K. Caldeira, and H. J. Schellnhuber, 2015: Long-term response of oceans to CO_2 removal from the atmosphere.* Nature Clim. Change, *doi:10.1038/nclimate2729.*
http://globalecology.stanford.edu/labs/caldeiralab/Caldeira_research/Mathesius_OceanResponse.html

[20] *Wikipedia, Coral*
https://en.wikipedia.org/wiki/Coral

[21] *NOAA, Earth System Research Laboratory, Recently Monthly Mauna Loa CO_2*
http://www.esrl.noaa.gov/gmd/ccgg/trends/

[22] *IPCC, 2013: Summary for Policymakers. In: Climate Change 2013: The Physical Science Basis. Contribution of Working Group I to the Fifth Assessment Report of the Intergovernmental Panel on Climate Change [Stocker, T.F., D. Qin, G.-K. Plattner, M. Tignor, S.K. Allen, J. Boschung, A.Nauels, Y.Xia, V.Bex and P.M. Midgley (eds.)]. Cambridge University Press, Cambridge, United Kingdom and New York, NY, USA.*
http://www.climatechange2013.org/images/report/WG1AR5_SPM_FINAL.pdf

[23] *Smithsonian National Museum of Natural History, Ocean Portal, Ocean Acidification*

The Ocean Portal Team; Reviewed by Jennifer Bennett (NOAA)
http://ocean.si.edu/ocean-acidification

[24] *Target atmospheric CO2: Where should humanity aim?*
J. Hansen (1 and 2), M. Sato (1 and 2), P. Kharecha (1 and 2), D. Beerling (3), R. Berner (4), V. Masson-Delmotte (5), M. Pagani (4), M. Raymo (6), D. L. Royer (7), J. C. Zachos (8) ((1) NASA GISS, (2) Columbia Univ. Earth Institute, (3) Univ. Sheffield, (4) Yale Univ., (5) LSCE/IPSL, (6) Boston Univ., (7) Wesleyan Univ., (8) Univ. California Santa Cruz)
http://arxiv.org/abs/0804.1126

[25] *Sandatlas, Basalt*
http://www.sandatlas.org/basalt/

[26] *Wikipedia, Carbonate*
https://en.wikipedia.org/wiki/Carbonate

[27] *Wikipedia, Calcium Carbonate*
https://en.wikipedia.org/wiki/Calcium_carbonate

[28] *NOAA, Ocean & Coasts Education Resources, Ocean Acidification*
http://www.noaa.gov/resource-collections/ocean-acidification

[28] *Cheaper Ways to Capture Carbon Dioxide, MIT Technology Review*
https://www.technologyreview.com/s/515881/cheaper-ways-to-capture-carbon-dioxide/

[29] *Carbon Dioxide Capture from Atmospheric Air Using Sodium Hydroxide Spray*
Joshuah K. Stolaroff,§ David W. Keith, ‡ and Gregory V. Lowry*,† Chemical and Petroleum Engineering, University of Calgary, and Departments of Civil and Environmental Engineering and Engineering and Public Policy, Carnegie Mellon University, Pittsburgh, Pennsylvania 15213
http://scholar.harvard.edu/files/davidkeith/files/97.stolaroff.aircapturecontactor.e.pdf

[30] *Large Scale CCS Projects, Global CCS Institute*
http://www.globalccsinstitute.com/projects/large-scale-ccs-projects

[31] *Global Status of CCS: 2014, Global CCS Institute*
http://hub.globalccsinstitute.com/sites/default/files/publications/180933/global-status-ccs-2014-supplementary-information-presentation-package.pdf

[32] http://web.stanford.edu/group/efmh/jacobson/Articles/I/CountryGraphs/TimelineWorld.jpg

[33] *Drinking-water, Fact sheet N°391, June 2015, World Health Organization*
http://www.who.int/mediacentre/factsheets/fs391/en/

[34] *Water scarcity, United Nations Website*
http://www.un.org/waterforlifedecade/scarcity.shtml

[35] *Water and Conflict, Events, Trends, and Analysis (2011-2012), Peter H. Gleick and Matthew Heberger.*
http://worldwater.org/wp-content/uploads/2013/07/www8-water-conflict-events-trends-analysis.pdf

[36] *Peak Water, Meena Palaniappan and Peter H. Gleick*
http://worldwater.org/wp-content/uploads/2013/07/ch01.pdf

[37] *Key world energy statistics 2016, International Energy Agency*
https://www.iea.org/publications/freepublications/publication/KeyWorld_Statistics_2015.pdf

[38] *Water-Energy Connection, United States Environmental Protection Agency*
https://www3.epa.gov/region9/waterinfrastructure/waterenergy.html

[39] Chapter two, The use of water today, World Water Council
www.worldwatercouncil.org/fileadmin/wwc/Library/WWVision/Chapter2.pdf

[40] Chapter two, The use of water today, World Water Council
www.worldwatercouncil.org/fileadmin/wwc/Library/WWVision/Chapter2.pdf

[41] *What is the minimum quantity of water needed? Water sanitation hygiene,* World Health Organisation
www.who.int/water_sanitation_health/emergencies/qa/emergencies_qa5/en/

[42] *Water, Facts and Trends, The World Business Council for Sustainable Development*
http://www.unwater.org/downloads/Water_facts_and_trends.pdf

[43] Igor A. Shiklomanov, State Hydrological Institute (SHI, St. Petersburg) and United Nations Educational, Scientific, and Cultural Organisation (UNESCO, Paris, 1999.
http://www.unep.org/dewa/vitalwater/jpg/0211-withdrawcons-sector-EN.jpg

[44] *Megascale Desalination, The world's largest and cheapest reverse-osmosis desalination plant is up and running in Israel, By David Talbot, MIT Technology Review*
https://www.technologyreview.com/s/534996/megascale-desalination/

[45] *Desalination out of Desperation, Severe droughts are forcing researchers to rethink how technology can increase the supply of fresh water, by David Talbot, December 2016,2014. MIT Technology Review*
https://www.technologyreview.com/s/533446/desalination-out-of-desperation

[46] *FAQ About Desalination, SIDEM Veolia*
http://www.sidem-desalination.com/Process/FAQ/

[47] *E20/435 Solar Panel Data Sheet, Sunpower*
https://us.sunpower.com/sites/sunpower/files/media-library/data-sheets/ds-e20-series-435-solar-panel-datasheet.pdf

[48] *Quadrennial Technology Review, an assessment of energy technology and research opportunities, September 2015*
http://energy.gov/sites/prod/files/2015/09/f26/Quadrennial-Technology-Review-2015_0.pdf

[49] *Life Cycle Assessment of utility-scale CDTE PV Balance of systems*, Parikhit Sinha, Mariska de Wild-Scholten, SmartGreenScans
http://smartgreenscans.nl/publications/Sinha-and-deWildScholten-2012-Life-cycle-assessment-of-utility-scale-CdTe-PV-Balance-of-Systems.pdf

[50] Wilburn, D.R., 2011, *Wind energy in the United States and materials required for the land-based wind turbine industry from 2010 through 2030: U.S. Geological Survey Scientific Investigations Report 2011–5036*, 22 p.
http://pubs.usgs.gov/sir/2011/5036/sir2011-5036.pdf

[51] *The carbon dioxide footprint of offshore wind: A Country Comparison.* F. Hinrichs, J. Goldsmith, C. Haro, S. Niaparast. Kungliga Tekniska Högskolan, MJ2413 Energy and Environment course, Autumn 2010
https://www.academia.edu/4298282/The_Carbon_Dioxide_Footprint_of_Offshore_Wind-_A_Country_Comparison

[52] *Renewable Energy's Hidden Costs, Low-carbon power depends on climate-unfriendly metal*, By John Matson, October 1, 2013. Scientific American.
http://www.scientificamerican.com/article/renewable-energys-hidden-costs/

[53] *Critical Materials for Sustainable Energy Applications*, Resnick Institute Report, September 2011
http://resnick.caltech.edu/docs/R_Critical.pdf

[54] *Copper: Essential in PV Solar Power Growth*, Copper Development Association inc.
https://www.copper.org/environment/sustainable-energy/pdf/CDA-Solar-Infographic.pdf

[55] *Copper's role in grid energy storage application*, Copper Development Association Inc.
https://www.copper.org/environment/sustainable-energy/pdf/CDA-Energy-Storage-Infographic.pdf

[56] *Copper's role in wind generation*, Copper Development Association Inc.
https://www.copper.org/publications/pub_list/pdf/a6179-Wind-Infographic.pdf

[57] *This commodity is going to be a huge beneficiary of the shift to electric cars. By Frank Holmes, April 19. 2016. Business Insider*
http://www.businessinsider.com/copper-to-benefit-from-electric-cars-2016-4

[58] *How much Copper does that electric car need? Mining Focus*
http://www.miningrocks.net/pdfs/copper.pdf

[59] *Rare minerals like Copper are critical to our economy, green energy, and technology. Copper Matters*
http://www.coppermatters.org/copper-powers-electric-cars/

[60] *Karen Smith Stegen (2015): Heavy rare earths, permanent magnets, and renewable energies: An imminent crisis. Energy Policy, 79, 1-8.*
http://www.jacobs-university.de/sites/default/files/downloads/heavy_rare_earths_permanent_magnets_and_renewable_energies_smith_stegen_2015.pdf

[61] *Zhang, Y., 2013: Peak Neodymium - Material Constraints for Future Wind Power Development. Master thesis in Sustainable Development at Uppsala University, No. 149, 41 pp*
https://www.diva-portal.org/smash/get/diva2:668091/FULLTEXT01.pdf

[62] *Scarce supply - the world's biggest rare earth metal producers, July 3. 2014. Mining-technology.com*
http://www.mining-technology.com/features/featurescarce-supply---the-worlds-biggest-rare-earth-metal-producers-4298126/

[63] *Rare Earth Metals: Will we have enough? By Renee Cho, September 19. 2012. Earth Institute, Columbia University*
http://blogs.ei.columbia.edu/2012/09/19/rare-earth-metals-will-we-have-enough/

[63] Rare Earth mining in China: the bleak social and environmental costs. By Jonathan Kaiman, March 20. 2014. The Guardian
https://www.theguardian.com/sustainable-business/rare-earth-mining-china-social-environmental-costs

[64] *Boom in mining rare earths poses mounting toxic risks. By Mike Ives, January 28. 2013. Yale, Environment 360*
http://e360.yale.edu/feature/boom_in_mining_rare_earths_poses_mounting_toxic_risks/2614/

[65] *5 facts to know about the California Methane Leak. By Tia Ghose, LiveScience, December 31. 2015*
http://www.scientificamerican.com/article/5-facts-to-know-about-the-california-methane-leak/

[65] http://web.stanford.edu/group/efmh/jacobson/Articles/I/AllCountries.xlsx

[66] *International Energy Outlook 2016, Chapter 5. Electricity. May 11. 2016. U.S. Energy Information Administration*
http://www.eia.gov/forecasts/ieo/electricity.cfm

[67] *Renewables 2016. Global Status Report, REN21*
http://www.ren21.net/wp-content/uploads/2016/06/GSR_2016_KeyFindings1.pdf

[68] *Energy, Electricity and Nuclear Power Estimates for the Period up to 2050, International Atomic Energy Agency. 2013 Edition*
http://www-pub.iaea.org/MTCD/Publications/PDF/RDS-1-33_web.pdf

[69] *Refining Silicon, PVEducation.org*
http://www.pveducation.org/pvcdrom/manufacturing/refining-silicon

[70] *Plans For New Reactors Worldwide, April 2016, World Nuclear Association*
http://www.world-nuclear.org/information-library/current-and-future-generation/plans-for-new-reactors-worldwide.aspx

[71] *The Low Dose, International Dose-Response Society*
http://dose-response.org/the-low-dose/

[72] *Coal and Gas are Far More Harmful than Nuclear Power, By Pushker Kharecha and James Hansen — April 2013*
http://www.giss.nasa.gov/research/briefs/kharecha_02/

[73] *Historical construction costs of global nuclear power reactors, Jessica R. Lovering, Arthur Yipa, Ted Nordhaus; The Breakthrough Institute, CA, USA; Department of Engineering and Public Policy, Carnegie Mellon University, PA, USA*
http://www.sciencedirect.com/science/article/pii/S0301421516300106

[74] *Advances in Small Modular Reactor Technology Development; A supplement to: IAEA Advanced Reactors Information System (ARIS) 2016 Edition*
https://aris.iaea.org/Publications/SMR-Book_2016.pdf

[75] *Update 1-China to build at least 60 nuclear plants in coming decade, Reuters, Fri Sept 16. 2016*
http://uk.reuters.com/article/china-nuclear-idUKL8N1BS3VB

[76] *SNC-Lavalin signs an agreement in principle for a Joint Venture with China National Nuclear Corporation & Shanghai Electric Company, September 22. 2016*
http://www.snclavalin.com/en/news/press-releases/2016/snc-lavalin-signs-agreement-principle-joint-venture-china-national-nuclear-corporation-shanghai-electric-company.aspx

[77] *Technical White Paper, July 2016, v2.0, Transatomic*
http://www.transatomicpower.com/wp-content/uploads/2015/04/TAP-White-Paper-v2.0.pdf

[78] *Lovering2016_OCC_Nuclear_Seven_Countries.xls*
https://drive.google.com/file/d/0Bzw8AHaXK2VYcEdEbGFMUW9iWXM/view

[79] *Nuclear Power Becomes Completely Renewable With Extraction Of Uranium From Seawater, October 3. 2016, by James Conca*
http://ansnuclearcafe.org/2016/10/03/nuclear-power-becomes-completely-renewable-with-extraction-of-uranium-from-seawater/#sthash.d11Y2iYL.8yHs5OYi.dpbs

[80] "Bryan, R.H., and I.T. Dudley. Estimated Quantities of Materials Contained in a 1000-MW(e) PWR Power Plant. ORNL-TM-4515, 1974".

[81] *"Peter Garrett & Klaus Rønde, "Life Cycle Assessment of Electricity Production from a V90-2.0MW Gridstreamer Wind Plant," Vestas Wind Systems A/S, December 2011. Table 5."*

[82] *ASME South African Workshop, AP1000 Overview, Ralph Hill*
http://files.asme.org/divisions/ned/16797.pdf

About the author

Mathijs was born in 1982, in Geleen, The Netherlands, and still lives there. He graduated from Hogeschool Zuyd in 2003 where he studied information technology. Before becoming a writer in 2012, Mathijs was independent in all his endeavors, he worked for multinational *Philips*, communication technology company *Royal KPN*, and pension fund investor *APG* (3rd largest in the world) as an IT specialist. Mathijs also owned a business for four years, again specializing in IT. In 2009 Mathijs started working on a new scouting method for a professional sports organization, which was based on defining metrics and performing statistical analyses, the culmination of which was a noticeable increase in attendance and overall scoring ability. In 2012, driven by a keen interest in science and technology, Mathijs decided to switch to writing about energy and climate change. His prime motivation for writing about these matters comes from a sense of responsibility, to leave this planet a better place than he found it.

Mathijs is an advocate for scientific thinking, skepticism, humanism, and the maximization of human potential.

Mathijs specializes in total sum energy calculations, communicating climate change science, and writing polemics and critiques on energy and science in politics.

In 2015 Mathijs published *Highway to Dystopia*—about climate change, energy, politics, economics, and religion; in 2016 he published *Science a la Carte*—about climate change and energy; in 2016 he published *The non-solutions project*—a rebuttal to the 100%WWS Roadmap by Mark Z. Jacobson.

Up until 2016, the year in which this book was published, Mathijs had no affiliations in related industries, research institutes or universities.

www.ingramcontent.com/pod-product-compliance
Lightning Source LLC
Chambersburg PA
CBHW071527180526
45170CB00010B/1475